Tau-*p*: A Plane Wave Approach to the Analysis of Seismic Data

MODERN APPROACHES IN GEOPHYSICS

formerly Seismology and Exploration Geophysics

VOLUME 8

The titles published in this series are listed at the end of this volume

Tau-*p*: a plane wave approach to the analysis of seismic data

Edited by
PAUL L. STOFFA

Department of Geological Sciences and the Institute for Geophysics,
The University of Texas at Austin, Austin, Texas, U.S.A.

Kluwer Academic Publishers
Dordrecht / Boston / London

Library of Congress Cataloging-in-Publication Data

Tau-p, a plane wave approach to the analysis of seismic
  data.

  (Modern approaches in geophysics)
  Includes index.
  1. Seismic prospecting. 2. Seismic waves.
I. Stoffa, Paul L., 1948-   II. Series.
TN269.T38 1989   622'.159   88-27363

Published by Kluwer Academic Publishers,
P.O. Box 17, 3300 AA Dordrecht, The Netherlands.

Kluwer Academic Publishers incorporates the publishing programmes of
D. Reidel, Martinus Nijhoff, Dr W. Junk and MTP Press.

Sold and distributed in the U.S.A. and Canada
by Kluwer Academic Publishers,
101 Philip Drive, Norwell, MA 02061, U.S.A.

In all other countries, sold and distributed
by Kluwer Academic Publishers Group,
P.O. Box 322, 3300 AH Dordrecht, The Netherlands.

ISBN-13: 978-94-010-6884-0   e-ISBN-13: 978-94-009-0881-9
DOI: 10.1007/978-94-009-0881-9

Contents

PAUL L. STOFFA

Introduction

In exploration seismology, data are acquired at multiple source and receiver positions along a profile line. These data are subsequently processed and interpreted.
The primary result of this process is a subsurface image of the exploration target.
As part of this procedure, additional information is also obtained about the
subsurface material properties, e.g., seismic velocities. The methods that are
employed in the acquisition and processing of exploration seismic data are
internally consistent. That is, principally near vertical incidence seismic waves are
generated, recorded and subsequently imaged. The data processing methods
commonly used are based upon a small angle of incidence approximation, thus
making the imaging problem tractable for existing data processing technology.
Although tremendously successful, the limitations of this method are generally
recognized.

Current and future exploration goals will likely require the use of additional
seismic waves, i.e., both compressional and shear precritical and postcritical
reflections and refractions. Also, in addition to making better use of seismic travel
times, recent efforts to directly incorporate seismic amplitude variations show that
the approach may lead to a better understanding of subsurface rock properties. In
response to more demanding exploration goals, recent data acquisition techniques
have improved significantly by increasing the spatial aperture and incorporating a
large number of closely spaced receivers. The need for better subsurface resolution
in depth and position has encouraged the use of 240, 512, and even 1024
recorded data channels with receiver separations of 5 to 25 m.

Because of the advent of supercomputers, this large volume of seismic data can
now be routinely processed. But, questions remain as to the best processing
procedures that should be applied to this data. Some of these questions are related
to the basic underlying assumptions used and others to the cost. Recognizing that
many new seismic analysis methods will likely be employed in the future, this book
emphasizes one promising approach to the processing and interpretation of
seismic data that has not yet been fully exploited, the analysis of seismic data after
a plane wave decomposition.

In the past, it was difficult to justify the use of plane wave techniques for
exploration reflection seismic data. Reflection seismic data were often sparsely
sampled with receiver separations of 50 and 25 m and maximum source receiver
offsets of little more than 3 km. In many cases, this degree of spatial sampling was

Paul L. Stoffa (ed.), Tau-p: A Plane Wave Approach to the Analysis of Seismic Data, vii—xii.
© 1989 *by Kluwer Academic Publishers.*

inadequate to perform a proper plane wave decomposition and the potential benefits of plane wave processing methods were not realized (although a result equivalent to conventional processing approaches could usually be achieved). Also, because only near-vertical incidence reflections were recorded, the ability to easily incorporate other seismic waves in the interpretation which is inherent in plane wave methods, was of little significance. Recent work that indicates the possibility of using seismic amplitude variations as indicators of lithologic differences, and the diagnostic properties of shear waves, make it necessary to investigate the use of other nonconventional processing and interpretation methods.

The papers in this book introduce the reader to the following concepts: plane wave decomposition; τ-p filtering methods that enhance or eliminate specific seismic waves, e.g., shear waves; velocity analysis methods in 1D, 2D or 3D that are not dependent upon the small angle of incidence approximation; and the direct inversion of plane wave seismic data. The common factors in the methods summarized are the analysis of seismic data in the domain of vertical delay or intercept time, τ, and horizontal ray parameter, p. The τ-p domain has been used conceptually by seismologists for the interpretation of seismic data for many years. What is now different is our ability to record seismic data with sufficient spatial aperture and sampling so that the recorded data can be directly transformed to this domain for processing and interpretation.

As the papers in this book imply, many limitations that are routinely encountered in conventional seismic processing and interpretation can be overcome by analyzing data in the τ-p domain. For example, for a 1D earth approximation, data recorded for any source receiver offset can be easily analyzed in the τ-p domain to determine interval velocities without the need for hyperbolic stacking or RMS velocities. Obviously, this has implications for direct inversion procedures that attempt to incorporate seismic data other than near-vertical incidence reflections. Also, the problem of estimating the source wavelet, critical to any inversion procedure can be directly addressed by analyzing plane wave data.

Unfortunately, the initial attempts to compute the τ-p transform for seismic data recorded as a function of source receiver offset and travel time, met with at least two difficulties. The first was the inadequate spatial sampling and the limited aperture of the routine data acquisition methods mentioned above. The second was the method of implementating the transform. Originally, the term 'slant stack' was routinely employed to describe the process of stacking seismic traces after applying a time delay to each seismic trace based on its source-receiver offset and the ray parameter seismogram being computed. Aliasing and windowing artifacts in the data transformed by this method, due to the limited spatial aperture and sampling, and the distortion of seismic pulse shapes and amplitudes made this procedure useful primarily for kinematic studies.

Recognition of the 'slant stack' process as part of the more general and fundamental process of a plane wave decomposition served to refine the method of implementation. In the first paper of this book, McCowan and Brysk, review past efforts to decompose point-source seismic data into plane wave components and the approximations employed. They then present a thorough theoretical description of the correct procedure for an axially symmetric earth structure based on cylindrical slant stacks. This is followed by examples which illustrate the differences between the original Cartesian slant stack and the geometrically correct cylindrical slant stack.

For the reader interested in computing the τ-p transform using cylindrical slant stacks, McCowan and Brysk include an appendix that outlines how to compute the cylindrical slant stack from Cartesian slant stacks. A computer program to compute the required weighting coefficients is also included.

The second paper in this volume illustrates the use of the τ-p domain and the transform procedure itself as a filtering method well suited to separate seismic events. In the paper by Tatham, the geometrically correct plane wave decomposition procedure is abandoned for a method based on the Cartesian slant stack which is modified to selectively improve the separation of different seismic events, e.g., compressional and shear reflections. By applying limits to the ray parameter values used in the transform, which vary with time based on regional stacking velocities, a time varying hyperbolic velocity filter is applied as part of the τ-p transform process. The detection and/or isolation of the targeted events in the τ-p domain is greatly enhanced by this filtering procedure. Further filtering to either remove or retain the seismic events is then accomplished by a straightforward windowing procedure in the τ-p domain. An example of the isolation and subsequent elimination of one type of seismic noise, i.e., ground roll, illustrates the method. Another problem, the attenuation of coherent scattered energy is attacked using the τ-p domain. Good results are obtained by time varying hyperbolic velocity filtering of shot records as part of the τ-p transform, followed by a simple windowing procedure in τ-p.

Tatham also addresses the problem of multiple attenuation. For a 1D earth structure multiple reflections are periodic in the τ-p domain and predictive deconvolution can be used to reduce their energy. In this case, the correct amplitudes of the plane wave data are required for the deconvolution procedure to work correctly. Consequently, Tatham employs the cylindrical slant stack procedure described by McCowan and Brysk in the first paper of this volume. The data are then deconvolved to attenuate the multiple energy. Rather than compute the inverse τ-p transform, Tatham corrects the τ-p data for normal move out (NMO) and stacks the resulting NMO corrected ray parameter traces directly in the τ-p domain. Comparison with conventional processing methods shows that the multiple energy has been significantly reduced.

The next paper in this volume discusses seismic travel times. Part of the original interest in the τ-p domain came from the recognition of the usefulness of the concept of intercept time for travel time inversion methods. Diebold begins with a derivation of the seismic travel time equation in three dimensions for iso-velocity but arbitrarily dipping layers. His derivation is based on a slowness decomposition of ray segments and is straightforward. The resulting travel time has a vertical delay time contribution, τ, in addition to contributions from the horizontal ray parameters in the x and y directions. The resulting travel time equation is succinct and exact.

Diebold's formulation of the 3D travel time equation is important for several reasons. First, with appropriate definitions, it is applicable to any source receiver geometry, including reversed 2D and 3D profiles, crossed and split-spread profiles, common mid-point profiles and buried sources and/or receivers. Second, it illustrates the fundamental nature of the vertical delay time contribution to seismic traveltimes. Once the contributions for the x and y delay times have been removed from the total travel time, the remaining component is the vertical delay time. Measurements of the horizontal delay time contributions are dependent only upon employing the proper data acquisition geometry and data sampling. These quantities can be directly derived from the field data and will be removed as part of the transform to the τ-p domain. The remaining quantity, the vertical delay time, τ, is the only quantity that depends on the earth structure.

Isolation of the vertical delay time contribution as the only part of seismic traveltime that is dependent upon the earth structure makes the fundamental role of τ apparent in problems of velocity analysis, imaging and inversion. Although these related problems are usually addressed by different formalisms and are characterized by diverse domains of implementation, they all implicitly require that the vertical delay time be either estimated or measured. For example, conventional CDP velocity analyses methods usually assume a 1D earth structure. The hyperbolic travel time assumption is equivalent to the single ellipse vertical delay time assumption in the τ-p domain. In either domain, NMO corrections are determined and the data NMO corrected. If the data are corrected in a consistent fashion for all offsets or ray parameters, the velocity used in the NMO correction is judged to be correct. But, in the τ-p domain, layer stripping which accounts properly for the ray bending in the overlying structure, and is exact for any subsurface angle of incidence, is just as easy to implement and use in the velocity analysis. In fact, this analysis can be done for any source or receiver geometry and for 2D or 3D earth strutures using the equations presented by Diebold.

Imaging of seismic data to depth requires that we migrate the seismic events from their observed travel time at each offset to their depth of reflection. In τ-p, this problem is changed to migrating the events from their vertical delay time at each horizontal ray parameter to depth. Since the vertical delay times of seismic arrivals contain all the information related to the earth structure, analysis of its form gives the velocity depth information we require in the imaging process. 1D, 2D and 3D reflection imaging of planar interfaces can be reduced to resampling the original plane wave decomposed seismograms.

Although Diebold does not specifically address the imaging problem, it is inherent in his discussion of τ-p travel times and the travel time inversion methods he outlines. He begins by reviewing 1D velocity inversion methods for precritical and postcritical reflections. He then addresses the limitations of each of the methods. Next, he reviews the classical dipping layer inversion procedure for refraction events and shows how precritical 2D reflection inversion can be accomplished. He concludes with a brief outline of how the τ-p travel time methods would be implemented in 3D, if the appropriate data were acquired.

The final two papers address the problem of inverting seismic data to directly determine subsurface material properties. The first paper by Tygel, Hubral, and Wenzel reviews plane wave composition and decomposition in terms of the Weyl integral. Both time harmonic and transient formulations are presented. The authors show how it is possible to obtain the source wavelet by exploiting the response of plane wave decomposed seismograms. No assumptions about the phase of the wavelet are required. By considering the theoretical properties of the plane wave reflection response, the regions where the amplitude spectrum of the wavelet and the phase spectrum of the wavelet can be recovered are identified. The authors point out that the seismic data required for this procedure, however, are the postcritical and inhomogeneous plane wave components. The methods proposed by the authors are theoretically justified, but data must be acquired that has sufficient spatial aperture so that the required data will be recorded and subsequently included in the plane wave decomposition.

The last paper in this book, by Ziolkowski *et al.*, describes an inversion procedure for common mid-point data after a plane wave decomposition. Using a 1D earth model, common mid-point data are first transformed into plane wave components. Assuming an acoustic stack of iso-velocity layers, a recursive formulation for the impulse response is inverted. The procedure described is a layer stripping approach and it is considered in the case of bandlimited data with additive noise. The authors emphasize the need for knowledge of the source wavelet for the procedure to be successful. They also discuss many practical problems, particularly the stabilization of the inversion. An outline that specifies each step in the inversion procedure is presented. The theoretical reasons for each step are described and the likely practical problems that will be encountered in the implementation are discussed.

Synthetic examples based on well log data are used to illustrate Ziolkowski *et al.*'s method. These examples are followed by tank data examples which show promising results. In both the synthetic and the tank data, the problem of including the low frequency information is discussed. The authors conclude with a real data example from the North Sea. After the inversion, synthetic seismograms were generated and found to compare favorably to the original data.

The papers in this volume were selected to illustrate several theoretical and practical aspects of analyzing data in the τ-p domain. Just as the Fourier transform of data from the time domain to the frequency domain serves to simplify many basic concepts in time series analysis, the plane wave decomposition of seismic data simplifies many of the basic concepts in seismic wave propagation. Further, just as the discrete Fourier transform offers in many cases a practical method of implementing fundamental filtering concepts, the plane wave decomposition of seismic data presents the opportunity to implement new and potentially more accurate methods of velocity analysis, imaging and inversion. Given the correct numerical implementation of the plane wave decomposition procedure, the only remaining fundamental problems to the practical exploitation of the theoretical concepts outlined by the authors in this book are the acquisition of properly sampled seismic data.

Department of Geological Sciences and the Institute for Geophysics,
The University of Texas at Austin,
8701 Mopac Blvd.,
Austin, TX 78759-8345,
U.S.A.

DOUGLAS W. McCOWAN and HENRY BRYSK

Cartesian and Cylindrical Slant Stacks

Abstract

The proper implementation of the $\tau-p$ method for surface data excited by a point source, requires the computation of a cylindrical slant stack. Usually, the common (Cartesian) slant stack is computed instead as an approximation to the geometrically-correct procedure. Here, we describe a formulation of the cylindrical slant stack as a weighted sum of Cartesian slant stacks which is accurate for all slownesses and efficient to perform.

As examples of the method, we show results computed from synthetic seismograms and from data recorded on the West Florida Shelf. Severe edge-effect noise which overwhelms the Cartesian slant stack, is attenuated by the cylindrical slant stack. Applications of the cylindrical slant stack to other seismological calculations, such as Lamb's problem, are also discussed. In particular, we show that the plane wave reflection coefficients apply exactly in the $\tau-p$ domain; hence an amplitude-versus-slowness analysis is unambiguous in this domain and requires no geometrical corrections to the data.

Introduction

We are primarily concerned with the theoretical aspects of applying the $\tau-p$ method to point-source data; however, we will first discuss the line-source geometry case for purposes of comparison. To simplify the nomenclature, we refer to the complete wave description (below the surface, as well as upon it) as the 'Radon transform' and to the superposition of observed traces on the surface as the 'slant stack', in both cases involving, for a point source, operations over two horizontal coordinates (Cartesian or cylindrical). To demonstrate the method, we successively consider a uniform medium, a vertically varying medium with axial symmetry, and a medium with arbitrary properties. We show that the slant stack can be carried out in the same fashion regardless of the medium, but the results for Radon transforms below the surface (i.e., the inversion problem) depend upon the properties of the medium.

The advantages of selective use of the vertical traveltime (τ) for seismic data reduction were detailed in the Russian literature (e.g., Bessonova *et al.*, 1974) before full computerization. The natural conjugate variable is the Snell parameter (p), or horizontal slowness, hence its designation as the $\tau-p$ method. However,

Paul L. Stoffa (ed.), Tau-p: A Plane Wave Approach to the Analysis of Seismic Data, 1—33.
© *1989 by Kluwer Academic Publishers.*

a number of authors have felt sufficiently uncomfortable with the Snell parameter that they resorted to a more cumbersome expression of the results in terms of the angle of incidence and the velocity separately. Stacking traces corresponding to the same p is usually referred to as a slant stack. A clear exposition of the $\tau-p$ approach was recently given by Diebold and Stoffa (1981) and Stoffa et al. (1981) who emphasized the geometrical concepts, the elimination of ground roll, and the separation of refractions, as well as applications to velocity analysis for extended spreads. Most development and practically all implementations of the slant stack have been restricted to treatment of a line source on the surface of a laterally-homogeneous earth. This can suffice for the retrieval of information determined from the relative phase (e.g., traveltime and raypaths). Calculations based on a two-dimensional (2-D) earth, however, do not correctly account for the wave amplitudes necessary for a full and accurate inversion of the data.

While the ray aspects of the $\tau-p$ interpretation were developed and implemented over many years, the corresponding wave aspects were practically ignored. Schultz and Claerbout (1978) first recognized the connection between slant stacking and the synthesis of plane-wave equivalents from the data. As Treitel et al. (1982) pointed out, the reflectivity method of wave analysis had been developed independently some time earlier, and closely related results on the synthesis of plane waves had appeared (e.g., Muller, 1971) without notice of the connection.

The most thorough development of the wave theory of the $\tau-p$ method was given by Chapman (1978, 1979, 1981). Chapman included explicit recognition of its proper application for point sources and for different geometries with their intrinsic symmetries, and also acknowledged its kinship to the much older mathematical formulation of the Radon transform. Chapman's presentation is concise and rigorous, exploring the mathematical ramifications but dealing tersely and intermittently with conceptual aspects. Henry et al. (1980), Phinney et al. (1981), and Harding (1985) concentrated on details of the numerical computations and related asymptotic evaluations. Treitel et al. (1982) covered wave propagation in a uniform medium. They emphasized the importance of the point-source wave treatment for optimal deconvolution. Furthermore, a number of papers applied the one-dimensional (1-D) slant stack to synthetic seismograms and to real data; they will not be enumerated here because they have no bearing on our discussion. The techniques and uses of the Radon transform in a more general context (with primary emphasis on medical applications) were detailed by Deans (1983) including a translation of Radon's 1917 paper and an extensive bibliography.

We illustrate the effect of cylindrical slant stacking on some wide-aperture marine seismic data. Specially, we compare a section obtained by the point-source treatment with the results of line-source slant stacking and its more common variants. The point-source slant stack was obtained by an efficient algorithm (which we describe in detail) developed for processing line-source slant stacks into a cylindrical slant stack.

Line Source

For a line source (taken as the y-axis) in a uniform medium, the acoustic equation reduces to

$$\frac{\partial^2 U}{\partial x^2} + \frac{\partial^2 U}{\partial z^2} - \frac{1}{c^2}\frac{\partial^2 U}{\partial t^2} = 0. \tag{1}$$

Its solution (subject to initial and boundary conditions) can be written in the form

$$U(t, x, z) = \int_{-\infty}^{\infty} d\omega \int_{-\infty}^{\infty} dp\, g(\omega, p)\, e^{-i\omega(t-px-qz)}, \tag{2}$$

with $p^2 + q^2 = c^{-2}$, as is easily verified by differentiation. In Equation (2), the exponential is a continuous 'plane' wave (propagating away from the y-axis) and $U(t, x, z)$ is expressed as a superposition of such waves. Each component wave has frequency ω and a Snell parameter p.

The usual slant stack is defined as

$$\psi(\tau, p') = \int_{-\infty}^{\infty} dx\, U(\tau + p'x, x, 0). \tag{3}$$

To clarify the meaning of the slant stack, we solve the reciprocal problem: Given the slant stacks $\psi(\tau, p')$, reconstruct $U(t, x, z)$. First, substitute Equation (2) into Equation (3),

$$\psi(\tau, p') = \int_{-\infty}^{\infty} d\omega \int_{-\infty}^{\infty} dp\, g(\omega, p)\, e^{-i\omega\tau} \int_{-\infty}^{\infty} dx\, e^{i\omega x(p-p')}$$

$$= 2\pi \int_{-\infty}^{\infty} \frac{d\omega}{\omega} \operatorname{sgn}(\omega)\, e^{-i\omega\tau} \int_{-\infty}^{\infty} dp\, g(\omega, p)\, \delta(p-p')$$

$$= 2\pi \int_{-\infty}^{\infty} \frac{d\omega}{\omega} \operatorname{sgn}(\omega)\, g(\omega, p')\, e^{-i\omega\tau}. \tag{4}$$

The factor $\operatorname{sgn}(\omega) = \omega/|\omega|$ arises from the change of variable in the innermost integral from x to ωx. For negative ω, ωx has the opposite sign from x, so the integration limits are reversed (equivalent to a sign change). On Fourier transforming back,

$$\int_{-\infty}^{\infty} d\tau\, \psi(\tau, p')\, e^{i\omega'\tau} = 4\pi^2 \operatorname{sgn}(\omega')\, g(\omega', p')/\omega' \tag{5}$$

so that the expansion coefficient g in Equation (2) is determined by the Fourier transform of the slant stack,

$$g(\omega, p) = \frac{1}{4\pi^2} \operatorname{sgn}(\omega)\,\omega \int_{-\infty}^{\infty} \psi(\tau, p)\, e^{i\omega\tau}\, d\tau. \tag{6}$$

Inversion of Equation (6) is complicated by the sgn (ω) factor which necessitates invoking the Hilbert transform. By definition, the Hilbert transform of ψ is

$$\Phi(\tau, p) = -\frac{1}{\pi} \int_{-\infty}^{\infty} \frac{\psi(t, p)\, dt}{\tau - t}, \tag{7}$$

or, more accurately, the Cauchy principal part of the integral. Taking the Fourier transform of Equation (7),

$$\int_{-\infty}^{\infty} d\tau\, e^{i\omega\tau}\, \Phi(\tau, p) = -\frac{1}{\pi} \int_{-\infty}^{\infty} dt\, \psi(t, p) \int_{-\infty}^{\infty} \frac{d\tau}{\tau - t}\, e^{i\omega\tau}, \tag{8}$$

and changing variables to $y = \omega(\tau - t)$ in the inner integral, we have

$$\int_{-\infty}^{\infty} \frac{d\tau}{\tau - t}\, e^{i\omega(\tau - t)} = \mathrm{sgn}(\omega) \int_{-\infty}^{\infty} \frac{dy}{y}\, e^{iy} = i\pi\, \mathrm{sgn}(\omega), \tag{9}$$

so that

$$\int_{-\infty}^{\infty} d\tau\, e^{i\omega\tau}\, \Phi(\tau, p) = -i\, \mathrm{sgn}(\omega) \int_{-\infty}^{\infty} dt\, e^{i\omega t}\, \Psi(t, p). \tag{10}$$

This relation between the Fourier transforms of a function and of its Hilbert transform is the basis of computer algorithms for calculating the Hilbert transform. After substituting Equation (10), the Fourier transform of Equation (6) is

$$\begin{aligned}
G(t, p) &= \frac{1}{2\pi} \int_{-\infty}^{\infty} d\omega\, g(\omega, p)\, e^{-i\omega t} \\
&= \frac{i}{8\pi^3} \int_{-\infty}^{\infty} d\omega\, \omega\, e^{-i\omega t} \int_{-\infty}^{\infty} d\tau\, \Phi(\tau, p)\, e^{i\omega\tau} \\
&= -\frac{1}{8\pi^3} \frac{d}{dt} \int_{-\infty}^{\infty} d\omega\, e^{-i\omega t} \int_{-\infty}^{\infty} d\tau\, e^{i\omega\tau \Phi(\tau, p)} \\
&= -\frac{1}{4\pi^2} \frac{d}{dt}\, \Phi(t, p).
\end{aligned} \tag{11}$$

Equation (11) supplies the ω integral in Equation (2),

$$U(t, x, z) = 2\pi \int_{-\infty}^{\infty} dp\, G(t - px - qz, p)$$

$$= -\frac{1}{2\pi} \frac{d}{dt} \int_{-\infty}^{\infty} dp\, \Phi(t - px - qz, p). \tag{12}$$

Equation (12), with z set to zero is, therefore, the reciprocal slant stack. Its expression in terms of the Hilbert transform appears in Chapman (1978, 1981). We obtained it in a different manner, to exhibit the 'plane-wave' decomposition.

Point Source in a Uniform Medium

For a point source acting at the origin of a uniform medium, the acoustic equation is

$$\frac{\partial^2 U}{\partial x^2} + \frac{\partial^2 U}{\partial y^2} + \frac{\partial^2 U}{\partial z^2} - \frac{1}{c^2}\frac{\partial^2 U}{\partial t^2} = 0. \tag{13}$$

Its solution has the form

$$U(t, x, y, z) = \int_{-\infty}^{\infty} d\omega \int_{-\infty}^{\infty} dp_x \int_{-\infty}^{\infty} dp_y\, g(\omega, p_x, p_y) \times$$

$$\times \exp[-i\omega(t-p_x x-p_y y-qz)], \tag{14}$$

with $p_x^2 + p_y^2 + q^2 = c^{-2}$. The exponential is now a true plane wave. The corresponding generalization of the slant stack is

$$\Psi(\tau, p_x', p_y') = \int_{-\infty}^{\infty} dx \int_{-\infty}^{\infty} dy\, U(\tau+p_x' x+p_y' y, x, y, 0). \tag{15}$$

Substituting Equation (14) into Equation (15), the same integrations are performed as in Equation (4), except now repeated for the pairs (x, p_x) and (y, p_y), yielding

$$\Psi(\tau, p_x', p_y') = 4\pi^2 \int_{-\infty}^{\infty} \frac{d\omega}{\omega^2}\, g(\omega, p_x', p_y')\, e^{-i\omega\tau}. \tag{16}$$

The two $\text{sgn}(\omega)$ factors have canceled. Inverse Fourier transformation of Equation (16) leads to

$$g(\omega, p_x, p_y) = \frac{1}{8\pi^3} \omega^2 \int_{-\infty}^{\infty} d\tau\, \Psi(t, p_x, p_y)\, e^{i\omega\tau}, \tag{17}$$

whose Fourier transform is obtained more directly as

$$G(t, p_x, p_y) = \frac{1}{2\pi} \int_{-\infty}^{\infty} d\omega\, g(\omega, p_x, p_y)\, e^{-i\omega t}$$

$$= \frac{1}{16\pi^4} \int_{-\infty}^{\infty} d\omega\, \omega^2\, e^{-i\omega t} \int_{-\infty}^{\infty} d\tau\, e^{i\omega\tau}\, \Psi(\tau, p_x, p_y)$$

$$= -\frac{1}{8\pi^3} \frac{d^2}{dt^2}\, \Psi(t, p_x, p_y). \tag{18}$$

This reciprocal slant-stack operation is therefore

$$U(t, x, y, z) = 2\pi \int_{-\infty}^{\infty} \mathrm{d}p_x \int_{-\infty}^{\infty} \mathrm{d}p_y \, G(t-p_x x-p_y y-qz, p_x, p_y)$$

$$= -\frac{1}{4\pi^2} \frac{\mathrm{d}^2}{\mathrm{d}t^2} \int_{-\infty}^{\infty} \mathrm{d}p_x \int_{-\infty}^{\infty} \mathrm{d}p_y \, \times$$

$$\times \, \Psi(t-p_x x-p_y y-qz, p_x, p_y). \tag{19}$$

The last result is also found in Chapman (1981), though derived somewhat differently.

Point Source in a Vertically Varying Axisymmetric Medium

We have already demonstrated the connection in a uniform medium between the slant stack and expansion of acoustic waves into plane waves. For simplicity, the discussion has been phrased exclusively in terms of propagating waves. This does not mean that evanescent waves are left out. The range of the p integrations in Equations (2) and (14) is infinite. When p^2 exceeds c^{-2}, q^2 is constrained to be negative and the z-dependent factor in the integrands is a real exponential (which, of course, must be chosen with negative exponent). If p is written as $p = \sin \Theta/c$, where Θ is the angle of incidence of the ray at the surface, Θ is then a complex number. If Θ is used as the variable instead of p, contour integration is required. The mathematical formulation is considerably more straight-forward in terms of p. The evanescent waves are there; we just have not discussed them.

We started with the simple case of a uniform medium to illustrate the basic concepts. The mathematical formulation survives practically intact for a laterally homogeneous medium whose velocity and density are allowed to vary with depth. As long as the velocity and density are functions of z only, separation of variables can be carried out for the more general acoustic equation, and the solution can be expressed as in Equations (2) and (14) with the single change that the factor $e^{i\omega qz}$ is replaced by some unspecified function $F(z, \omega, p)$ which depends upon the velocity and density profiles. The function can be normalized (without loss of generality), so that $F(0, \omega, p) = 1$. The calculations of the slant stack remain unaffected, because they involve only the values of U on the surface. Similarly, the reciprocal slant stack is unaltered (again only surface values are used). The description of the waves on the surface is unchanged; the waves are, however, distorted since the properties of the medium change below the surface. The rays are loci of the constant Snell parameter p.

For problems in which lateral homogeneity has been replaced by some other symmetry for which the acoustic equation is separable, parallel development is possible in the appropriate coordinate system. Chapman (1978, 1979, 1981) developed the acoustic equation in cylindrical and spherical geometry, deriving the counterparts of Equations (31)–(41).

Point Source in Cylindrical Coordinates

A drawback to the current development of the point-source problem thus far is that the slant stack is expressed, not in terms of the radial Snell parameter p, but rather in terms of its orthogonal components p_x and p_y. This problem is easily remedied by transcribing Equations (15) and (19) to cylindrical coordinates, both in space and in the p variables, setting

$$(x, y) = (r \cos \phi, r \sin \phi),$$

and   (20)

$$(p_x, p_y) = (p \cos \alpha, p \sin \alpha).$$

Retaining the same symbols U and Ψ for these variables reformulated as functions of the cylindrical coordinates, Equations (15) and (19) read

$$\Psi(\tau, p, \alpha) = \int_0^\infty r \, dr \int_{-\pi}^\pi d\phi \, U[\tau + pr \cos(\phi - \alpha), r, \phi, 0],$$ (21)

and

$$U(t, r, \phi, z) = -\frac{1}{4\pi^2} \frac{d^2}{dt^2} \int_0^\infty p \, dp \int_{-\pi}^\pi d\alpha \times$$

$$\times \Psi[t - pr \cos(\phi - \alpha) - qz, p, \alpha].$$ (22)

If the problems is fully axisymmetric, i.e., in the properties of the source, as well as in the medium, the inherent symmetry dictates that U and Ψ should be independent of angle. Specifically, the slant stack should vary with the magnitude of p, Snell's parameter, but not with the direction of propagation. However, the angular integrations in Equations (21) and (22) are not eliminated. These integrations involve the relative orientation of the position vector and of the direction of propagation.

Equation (21) defines a 2-D slant stack over surface data and implies areal coverage in the acquisition, i.e., data not only at different offsets r, but also in all directions ϕ. The ϕ integral in Equation (21) extends over a ring (on the surface) of radius r about the source; it involves stacking the traces located on the ring (with interpolation as necessary) with a different time shift applied to each trace depending upon its azimuth. For a given slant stack (fixed direction of propagation α and Snell's parameter p), the time shift is the product of p and the projection of the position vector (of magnitude r) along the direction α. If the acquisition coverage consists of a single line (along which the source also lies), real data are available either at one point on the ring or at two points 180 degrees apart. In an axisymmetric problem, the data are expected to depend upon offset but not upon direction, so that

$$U(t, r, \phi, z) = U(t, r, 0, z) = U(t, r, \alpha, z).$$ (23)

Note, however, that the expression for the argument t in the integrand of

Equation (21) retains its azimuthal dependence. If axial symmetry is assumed, we can imagine the existing trace at offset r (or the average of the two existing traces) to be repeated at every ϕ grid point on the ring. The ϕ integration then consists of stacking multiple copies of the existing trace at offset r, each time shifted a different amount according to the ϕ value assigned to it. Whether areal coverage is real or simulated, the r integration consists simply of stacking the stacked traces obtained at each offset again by the ϕ integral procedure, where each stacked trace is multiplied by the offset.

The procedure just outlined permits us to relate the cylindrical slant stack directly to the line-source slant stack when there is axial symmetry. Substitute Equation (23) into Equation (21); take $\phi' = \phi - \alpha$ as the integration variable instead of ϕ. Using the symmetry properties of the cosine, Equation (21) can then be rewritten as

$$\Psi(\tau, p, \alpha) = 2 \int_0^\infty r \, dr \int_0^{\pi/2} d\phi' [U(\tau + pr \cos \phi', r, 0, 0) +$$
$$+ U(\tau - pr \cos \phi', r, 0, 0)]. \tag{24}$$

Interchange the order of integration. In the first term, let $r = x$; in the second term, let $r = -x$. Taking

$$U(t, x, 0, 0) = U(t, -x, 0, 0), \tag{25}$$

as mandated by axial symmetry, Equation (24) can be recast into

$$\Psi(\tau, p, \alpha) = 2 \int_0^{\pi/2} d\phi' \int_{-\infty}^\infty dx \, |x| \, U(\tau + px \cos \phi', x, 0, 0). \tag{26}$$

The ϕ' integration is trivial for $p = 0$,

$$\Psi(t, 0, \alpha) = \pi \int_{-\infty}^\infty dx \, |x| \, U(t, x, 0, 0). \tag{27}$$

For $p > 0$, change variables from ϕ' to $p' = p \cos \phi'$. Then

$$dp' = -p \sin \phi' \, d\phi' = -(p^2 - p'^2)^{1/2} \, d\phi', \tag{28}$$

and

$$\Psi(\tau, p, \alpha) = 2 \int_0^p dp' \, (p^2 - p'^2)^{-1/2} \times$$
$$\times \int_{-\infty}^\infty dx \, |x| \, U(\tau + p'x, x, 0, 0). \tag{29}$$

It is useful to reinterpret Equations (27) and (29) by transposing the concepts just mentioned. Consider a seismic line along the x-axis, with a shotpoint at $x = 0$. Multiply each trace by the magnitude of its offset. Perform a conventional slant stack upon the rescaled traces (appropriate for a line source along the y-axis). The result is precisely the x integral in Equations (27) and (29), for Snell's parameter p'. [The reinterpretation of U in Cartesian coordinates is legitimate: x is r or $-r$, and the sign is no problem in view of Equation (25); the next to last argument of U remains 0 when read as y.] The outer integral of Equation (29) then expresses the cylindrical slant stack for p as a superposition of the line-source slant stacks (using traces to which linear offset gain has been applied) for all p' close to p. From Equation (24), note that its first term corresponds to the positive x values in Equations (27) and (29); its second term corresponds to the negative values. The implied assumption is that the data include a full set of receiver locations on both sides of the shotpoint. If the spread is actually one-sided, the contribution corresponding to the second term in Equation (24) must be supplied by negative Snell's parameters. That is,

$$\Psi(\tau, p, \alpha) = 2 \int_0^p dp' \, (p^2 - p'^2)^{-1/2} \int_0^\infty r \, dr \times$$
$$\times \, [U(\tau + p'r, r, 0, 0) + U(\tau - p'r, r, 0, 0)], \tag{30}$$

with the line-source slant stacks for p' and $-p'$ combined. Numerical evaluation of the integral is discussed is detail in Appendix A and a computer program to evaluate it is given in Appendix B.

When axial symmetry is assumed, the shuffling of traces described can be traded off against more elaborate integrations. Equation (21) can be recast as

$$\Psi(\tau, p, \alpha) = \int_0^\infty r \, dr \int_{-\pi}^\pi d\phi \int_{-\infty}^\infty dt \, U(t, r, \phi, 0) \times$$
$$\times \, \delta[t - \tau - pr \cos(\phi - \alpha)]$$
$$= \frac{1}{2\pi} \int_0^\infty r \, dr \int_{-\pi}^\pi d\phi \int_{-\infty}^\infty dt \, U(t, r, \phi, 0) \times$$
$$\times \int_{-\infty}^\infty d\omega \, \exp\{i\omega[t - \tau - pr \cos(\phi - \alpha)]\}. \tag{31}$$

For a axisymmetric problem, the value of U is independent of ϕ. The remaining ϕ integration is just the integral representation of the Bessel function

$$\frac{1}{2\pi} \int_{-\pi}^\pi d\phi \, \exp[-i\omega pr \cos(\phi - \alpha)] = J_0(\omega pr). \tag{32}$$

Simplifying the notation by setting $U(t, r) = U(t, r, \phi, 0)$, dropping the argument α from Ψ (since the value of Ψ is independent of α), and reordering the integrations, Equation (31) becomes

$$\Psi(\tau, p) = \int_{-\infty}^\infty d\omega \, e^{-i\omega\tau} \int_0^\infty r \, dr \, J_0(\omega pr) \int_{-\infty}^\infty dt \, e^{i\omega t} \, U(t, r). \tag{33}$$

Equation (33) consists of a Fourier transform of the data, followed by integration with a Bessel function, then a Fourier transform back. For $p = 0$, the Bessel function reduces to unity and the Fourier transformations cancel, leaving

$$\Psi(\tau, 0) = 2\pi \int_0^\infty r \, dr \, U(\tau, r). \tag{34}$$

For $p > 0$, the Fourier transform of the Bessel function is

$$\int_{-\infty}^\infty d\omega \, J_0(\omega pr) \, e^{i\omega(t-\tau)}$$

$$= 2[p^2r^2 - (t-\tau)^2]^{-1/2}, \quad \text{if } pr > |t-\tau|,$$

$$= 0, \qquad\qquad\qquad \text{if } pr < |t-\tau|. \tag{35}$$

For given r, Equation (35) establishes a finite range of t in which the integral does not vanish, so that

$$\Psi(\tau, p) = 2 \int_0^\infty r \, dr \int_{\tau-pr}^{\tau+pr} dt [p^2r^2 - (t-\tau)^2]^{-1/2} \, U(t, r). \tag{36}$$

Results substantially equivalent to Equations (31)–(36) were obtained by Chapman (1978, 1981). Chapman also discusses an approximation to the cylindrical slant stack valid when pr is large compared to the periods of interest which is useful in refraction data analysis. While the three integrations in Equation (33) have been reduced to two, the t integral has integrable singularities at both ends, which are obstacles to numerical integration. These singularities can be eliminated by an integration by parts. There is an advantage in first changing variables, from t to $T = |t-\tau|$. Then

$$\Psi(\tau, p) = 2 \int_0^\infty r \, dr \int_0^{pr} dT (p^2r^2 - T^2)^{-1/2} \times$$

$$\times [U(\tau+T, r) + U(\tau-T, r)]. \tag{37}$$

Note that Equation (37) is equivalent to Equation (30). The integration by parts in T yields

$$\Psi(\tau, p) = 2 \int_0^\infty r \, dr \int_0^{pr} dT \arccos \frac{T}{pr} \times$$

$$\times \left[\frac{dU(\tau+T, r)}{dT} + \frac{dU(\tau-T, r)}{dT} \right]. \tag{38}$$

The derivative can be extracted from the integrals by using

$$\frac{d}{dT} U(\tau \pm T, r) = \pm \frac{d}{d\tau} U(\tau \pm T, r). \tag{39}$$

Returning to t as the variable,

$$\Psi(\tau, p) = 2\frac{d}{d\tau}\int_0^\infty r\, dr \int_{\tau-pr}^{\tau+pr} dt \times$$

$$\times \operatorname{sgn}(t-\tau)\arccos\frac{|t-\tau|}{pr}\, U(t, r). \tag{40}$$

A straightforward alternate procedure is to interchange the order of integrations in Equation (36) or (37), integrate by parts in r, then switch back. The result is

$$\Psi(\tau, p) = -\frac{2}{p^2}\int_0^\infty dr \int_{\tau-pr}^{\tau+pr} dt \times$$

$$\times [p^2r^2 - (t-\tau)^2]^{1/2}\frac{dU(t, r)}{dr} \tag{41}$$

which looks simpler than Equation (40), but leaves the derivative operation in the integrals. With axial symmetry, Equations (21), (29), (33), (40), and (41) represent alternative approaches to the evaluation of the slant stack, involving various tradeoffs. The practical choice among them depends on the available computer system.

The reciprocal slant stack is given by Equation (22) with z set to zero. The integrals are exactly the same form as in Equation (21), thus their evaluation in the axisymmetric case can be directly transcribed from the last paragraph. In the counterpart of Equation (33), the double time derivative can be performed by inserting $(-\omega^2)$ into the integrand.

Recently, Harding (1985) presented cylindrical slant stacks computed using Equation (33). While his results are equivalent to ours [based on Equation (30)], his procedure is not as amenable to implementation on an array processor. Slant stacking (actually beam forming) is an inherently simple procedure with many repetitive steps that can be efficiently programmed on an array processor.

Table I is a summary of slant stacks and their inverses for both Cartesian and cylindrical geometry. The cylindrical case is simplest when expressed in the frequency domain. The alternative involves convolution with the Fourier transform of the J_0 Bessel function and can be found in Chapman (1981). Note that in the Cartesian case that, when U is symmetric with respect to x, Ψ will be symmetric with respect to p. The integrals should still both be performed over the complete range from $-\infty$ to $+\infty$.

Point Source in an Arbitrary Medium

Formally, slant stacking can be applied to data distributed over an areal coverage without assuming any geometric symmetry. The relationship between the slant

Table I. Slant stack pairs

Geometry	Forward	Inverse
Cartesian	$\Psi(\tau,p) = \int_{-\infty}^{+\infty} \mathrm{d}x\, U(\tau+px, x)$	$U(t,x) = -\dfrac{1}{2\pi}\int_{-\infty}^{+\infty} \mathrm{d}p\, \Psi^{\dagger}(t-px, x)$
Cylindrical	$\hat{\Psi}(\omega,p) = \int_{0}^{\infty} r\,\mathrm{d}r \int_{-\infty}^{+\infty} \mathrm{d}t\, e^{i\omega t} J_0(\omega pr)\, U(t,r)$ $\Psi(\tau,p) = \dfrac{1}{2\pi}\int_{-\infty}^{+\infty} \mathrm{d}\omega\, e^{-i\omega t} \hat{\Psi}(\omega,p)$	$U(t,r) = -\dfrac{1}{2\pi}\dfrac{\mathrm{d}^2}{\mathrm{d}t^2}\int_{0}^{\infty} p\,\mathrm{d}p \int_{-\infty}^{+\infty} \mathrm{d}\omega\, e^{-i\omega t} J_0(\omega pr)\, \hat{\Psi}(\omega,p)$

$\dagger \equiv$ time derivative of Hilbert transform.

stack and its reciprocal is retained. Again, this is most easily demonstrated in Cartesian coordinates. Starting with the counterpart of Equation (31),

$$\Psi(\tau, p_x, p_y) = \int_{-\infty}^{\infty} dx \int_{-\infty}^{\infty} dy \, U(\tau + p_x x + p_y y, x, y, 0)$$

$$= \int_{-\infty}^{\infty} dx \int_{-\infty}^{\infty} dy \int_{-\infty}^{\infty} dt \, U(t, x, y, 0) \times$$

$$\times \, \delta(t - \tau - p_x x - p_y y)$$

$$= \frac{1}{2\pi} \int_{-\infty}^{\infty} dx \int_{-\infty}^{\infty} dy \int_{-\infty}^{\infty} dt \, U(t, x, y, 0) \times$$

$$\times \int_{-\infty}^{\infty} d\omega \exp[i\omega(t - \tau - p_x x - p_y y)]$$

$$= \frac{1}{2\pi} \int_{-\infty}^{\infty} d\omega \, e^{-i\omega\tau} \int_{-\infty}^{\infty} dx \, e^{-i\omega p_x x} \int_{-\infty}^{\infty} dy \, e^{-i\omega p_y y} \times$$

$$\times \int_{-\infty}^{\infty} dt \, e^{i\omega t} \, U(t, x, y, 0). \tag{42}$$

In the last step, the equation is restructured as a sequence of Fourier transforms. The reciprocal operation is achieved by Fourier transforming back four times. The only subtlety is that the transform variable for x is ωp_x and for y it is ωp_y, so that an extra factor of ω^2 appears. U thus has the form

$$U(t, x, y, 0) = \frac{1}{8\pi^3} \int_{-\infty}^{\infty} d\omega \, e^{-i\omega t} \, \omega^2 \int_{-\infty}^{\infty} dp_x \, e^{i\omega p_x x} \times$$

$$\times \int_{-\infty}^{\infty} dp_y \, e^{i\omega p_y y} \int_{-\infty}^{\infty} d\tau \, e^{i\omega\tau} \Psi(\tau, p_x, p_y)$$

$$= -\frac{1}{8\pi^3} \frac{d^2}{dt^2} \int_{-\infty}^{\infty} d\omega \, e^{-i\omega t} \int_{-\infty}^{\infty} dp_x \, e^{i\omega p_x x} \times$$

$$\times \int_{-\infty}^{\infty} dp_y \, e^{i\omega p_y y} \int_{-\infty}^{\infty} d\tau \, e^{i\omega\tau} \Psi(\tau, p_x, p_y). \tag{43}$$

We can move backward through the steps in Equation (42) to arrive at

$$U(t, x, y, 0) = -\frac{1}{4\pi^2} \frac{d^2}{dt^2} \int_{-\infty}^{\infty} dp_x \int_{-\infty}^{\infty} dp_y \times$$

$$\times \, \Psi(t - p_x x - p_y y, p_x, p_y). \tag{44}$$

The slant stack can be performed either directly as a double integral over time-shifted traces, or by a four-fold sequence of Fourier transforms; so can its reciprocal, except for the differentiation.

Our earlier discussion concentrated on interpretation of the slant stack when there is an inherent symmetry for which the acoustic equation is separable. We have just seen that the slant stack and its reciprocal can be carried out as a sequence of Fourier transformations without postulating any symmetry. Further insight can be gained by closely examining the process of Fourier decomposition and reconstitution. With the inner three integrations in the last line of Equation (42), we expanded the pressure wave at the surface into plane waves and picked out the particular plane wave with a frequency ω and a wavenumber with x and y components ωp_x and ωp_y. With the last integration, we superpose a collection of such plane waves to build up a new wavelet. We select the constituent plane waves by the following rule: they all have the same components of the Snell's parameter p_x and p_y; hence they have the same direction of incidence into the medium, but different frequencies (and wavenumbers). Hence, the slant stack represents a wavelet entering the medium with a specific orientation. The process has been referred to as 'beam forming'. In the ω integration, we could skew the superposition by introducing some function of ω, an 'antenna pattern'. While we performed the exercise in Cartesian coordinates with x and y Fourier transforms, the procedure can easily be transcribed to cylindrical coordinates and Fourier–Bessel transforms.

Radon Transform

In a uniform medium, successive application of Equations (21) and (22) permits reconstruction of the acoustic waves at depth from their observed values at the surface. This is possible because in a uniform medium there are no reflections or refractions, only the wave spreading from the source. The (idealized) data consist of just the direct arrival, so it is not surprising that the wavefront can be continued below the surface. On the other hand, for a laterally homogeneous medium varying with depth (i.e., velocity c and density ρ both functions only of z), the z-dependent factor in Equation (22) is a function $F(z, \omega, p)$ which has to be determined by substitution into the full acoustic equation, leading to a differential equation for F as a function of z. The fundamental problem of seismic inversion is to apply the data to solving the acoustic equation (or some equivalent) for the velocity and density profiles — a completely different matter and a difficult one.

Thus far, the slant stack has been performed over the plane $z = 0$, described as the surface and also as the data plane. Actually, the mathematical operation is defined as a surface integral at fixed z, and it can be applied to any plane $z = $ constant. Indeed, the procedure can be extended to every z. The depth-dependent generalization of the slant stack (the Radon transform) is

$$S(\tau, p, z) = \int_0^\infty r\, dr \int_{-\pi}^{\pi} d\phi\, U(\tau + pr \cos \phi, r, \phi, z) \tag{45}$$

by extension of Equation (21), with α suppressed when there is axial symmetry.

The acoustic equation is then easily transcribed to a corresponding equation for S in the axisymmetric case. Working in Cartesian coordinates, the equation

$$\frac{\partial^2 U}{\partial x^2} + \frac{\partial^2 U}{\partial y^2} + \frac{\partial^2 U}{\partial z^2} - \frac{d \ln \rho(z)}{dz} \frac{\partial U}{\partial z} - \frac{1}{c^2(z)} \frac{\partial^2 U}{\partial t^2} = 0 \tag{46}$$

is Fourier transformed four times, as in Equation (42). The net result is

$$\frac{\partial^2 S}{\partial z^2} - \frac{d \ln \rho(z)}{dz} \frac{\partial S}{\partial z} - \left[\frac{1}{c^2(z)} - p^2 \right] \frac{\partial^2 S}{\partial \tau^2} = 0. \tag{47}$$

For $p = 0$, Equation (47) is identical to Equation (46) except that the 3-D Laplacian operator has been reduced to its 1-D equivalent. For any p, the problem in three spatial variables has been scaled down to a problem in one spatial variable, with time t replaced by intercept time τ and the inverse velocity replaced by the vertical component of the slowness. Stepping through Equation (42), a point source $\delta(x)\delta(y)\delta(z)f(t)$ is transformed to $\delta(z)f(\tau)$. In every respect, the point-source problem has been reduced to a plane-wave problem. This is the distinct advantage of attacking the acoustic inversion problem in the slant-stack formulation.

In the absence of axial symmetry, i.e., when ρ and c depend upon the horizontal coordinates, Fourier transforming the acoustic equation [the generalization of Equation (46)] no longer leads to a differential equation for the Radon transform such as Equation (47). The coefficients involving c and derivatives of ρ do not factor out the spatial Fourier transforms. Slant stacks can still be performed at different z levels, but we cannot expect to relate them by an equivalent acoustic equation for the Radon transform.

Lamb's Problem under Radon Transformation

Because the Radon transformation turns point-source data into an equivalent plane-wave problem, and because the concept and derivation of reflection co-efficients pertain to plane waves, the $\tau-p$ domain is the preferred representation for reflection amplitude studies. Consider the textbook Lamb's problem: homogeneous half-spaces excited by a point source. Aki and Richards's (1980) presentation is most convenient for our purposes because the problem is expressed in terms of p. The interface is the horizontal plane $z = 0$. The source and receiver are in the first medium at distances z_0 and z, respectively, from the interface and with a horizontal offset r. The z-axis is defined as positive into the second medium. In the acoustic case, an incident monochromatic spherical wave

$$U_{inc} = \frac{1}{R} e^{-i\omega(t - R/c_1)}, \tag{48}$$

where

$$R = [r^2+(z-z_0)^2]^{-1/2},$$

is expanded using the Sommerfeld integral as

$$U_{inc} = i\omega\, e^{-i\omega t} \int_0^\infty p'\, dp'\, (c_1^{-2}-p'^2)^{-1/2}\, J_0(\omega p'r) \times$$
$$\times \exp[i\omega(c_1^{-2}-p'^2)^{-1/2}|z-z_0|] \qquad (49)$$

and the corresponding reflected and transmitted waves are

$$U_{refl} = i\omega\, e^{-i\omega t} \int_0^\infty B(p')p'\, dp'\, (c_1^{-2}-p'^2)^{-1/2}\, J_0(\omega p'r) \times$$
$$\times \exp[-i\omega(c_1^{-2}-p'^2)^{1/2}(z+z_0)], \qquad (50)$$

and

$$U_{trans} = i\omega\, e^{-i\omega t} \int_0^\infty C(p')p'\, dp'\, (c_1^{-2}-p'^2)^{-1/2}\, J_0(\omega p'r) \times$$
$$\times \exp\{-i\omega[(c_1^{-2}-p'^2)^{1/2} z_0-(c_2^{-2}-p'^2)^{1/2} z]\}. \qquad (51)$$

Note that z_0 is negative, as is z in Equations (49) and (50), but not in Equation (51). After the coefficients B and C are determined from the boundary conditions, Aki and Richards (1980) evaluated the expression in Equation (50) as a contour integral by the method of steepest descent, yielding in the high-frequency limit (or more precisely, for the leading term in an asymptotic expansion in $c_1/\omega R_0$)

$$U_{refl} = \frac{B(p_s)}{R_0}\, e^{-\omega(t-R_0/c_1)}, \qquad (52)$$

where

$$R_0 = [r^2+(z+z_0)^2]^{1/2},$$

and where p_s is the value of p satisfying geometrical optics, i.e.,

$$c_1 p_s = \sin\Theta_s = r/R_0, \qquad \cos\Theta_s = |z+z_0|R_0, \qquad (53)$$

and

$$R_0 = |z+z_0|/(1-c_1^2 p_s^2)^{1/2}.$$

Brekhovskikh (1960) carried out the calculation in angular coordinates for both U_{refl} and U_{trans}. The treatment of elastic media follows the same pattern, as given explicitly by Aki and Richards (1980) for the $P-P$ reflection and by Brekhovskikh (1960) for all cases. Brekhovskikh (1960) also calculated the next order term for acoustic waves. The corresponding result for elastic waves was given by Krail and Brysk (1983).

When Equations (49)–(51) are Radon-transformed by application of Equation (33), the t and r integrations yield δ-functions, so that the multiple integrals collapse to

$$S_{\text{inc}} = \frac{2\pi i}{\omega}(c_1^{-2}-p^2)^{-1/2}\, e^{-i\omega\tau} \times$$
$$\times \exp[i\omega(c_1^{-2}-p^2)^{1/2}|z-z_0|], \tag{54}$$

$$S_{\text{refl}} = \frac{2\pi i}{\omega}(c_1^{-2}-p^2)^{-1/2}\, B(p)\, e^{-i\omega\tau} \times$$
$$\times \exp[-i\omega(c_1^{-2}-p^2)^{1/2}(z+z_0)], \tag{55}$$

and

$$S_{\text{trans}} = \frac{2\pi i}{\omega}(c_1^{-2}-p^2)^{-1/2}\, C(p)\, e^{i\omega\tau} \times$$
$$\times \exp\{-i\omega[c_1^{-2}-p^2)^{1/2}\, z_0-(c_2^{-2}-p^2)^{1/2}\, z]\}. \tag{56}$$

Equations (54)–(56) are plane waves; they are the exact solution in the $\tau-p$ domain to Lamb's problem as posed. The extension to elastic media is obvious. The reflection coefficients (or Zoeppritz coefficients) are the same in the $\tau-p$ domain as in ordinary spacetime. The propagation phases, on the other hand, are rescaled, with time t replaced by intercept time τ and the inverse velocity foreshortened to its vertical component, as might have been anticipated from comparison of Equations (46) and (47).

Some Computational Examples

A comparison between the Cartesian and cylindrical slant stacks can be easily seen when they are applied to synthetic data. Figure 1 shows a $\tau-p$ synthetic seismogram computed by the reflectivity method (e.g., Fryer, 1980) from the model parameters given in Table II. The seismogram includes P to S conversions and all interbed multiples, but no free surface reflections. The plot is of the vertical component of displacement at the free surface, which accounts for the lack of direct arrivals. The model is typical of an oceanic refraction experiment, including low velocity zones in both P and S wave velocities.

Cartesian and cylindrical slant stacks of the model are shown in Figures 2 and 3, respectively. These were computed from the inverse formulas given in Equations (12) and (22), respectively. In the first case, Figure 2 represents a synthetic seismogram in plane geometry excited by an infinite line source. Conversely, Figure 3 represents the same in three-dimensional, cylindrically symmetric geometry excited by a point source. The most obvious differences between these two examples are in the asymptotic dropoffs in time and space. This is because a line source of radiation produces waves that decay as $1/\sqrt{r}$ at large distances. How-

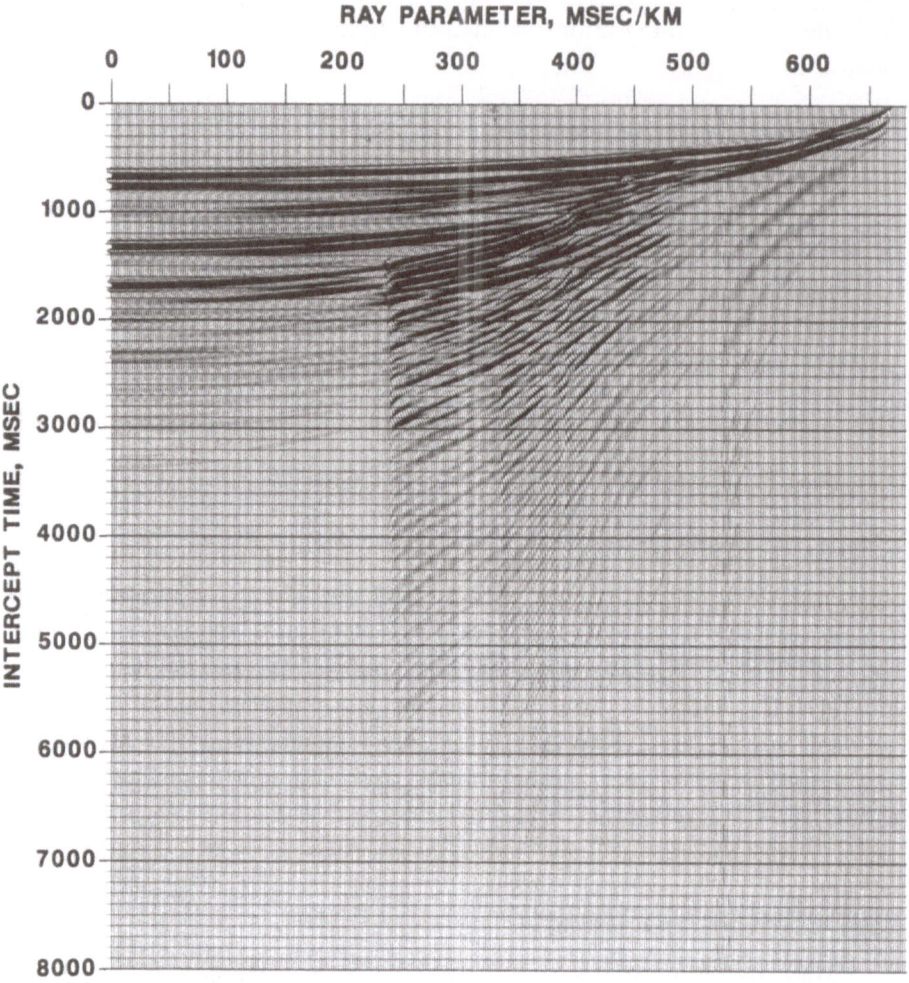

RAY PARAMETER, MSEC/KM

Fig. 1. τ–p synthetic seismogram for the model given in Table II. The model includes P-waves and their S-wave conversions with all interbed multiples but no free surface reflections. Note that the interbed multiples are efficiently generated near the layer critical points.

Table II.   Model parameters

Thickness (m)	p velocity (m/sec)	s velocity (m/sec)	Density (gm/cc)	P Wave Q	S Wave Q	2-way time (msec)
500	1500	0	1.0	∞	0	0
100	2000	1300	1.7	200	20	667
250	2200	2000	1.9	250	25	767
300	1800	1400	2.2	300	30	994
450	2500	1800	2.9	500	345	1327
∞	4200	3000	2.9	500	340	1687

OFFSET, METERS

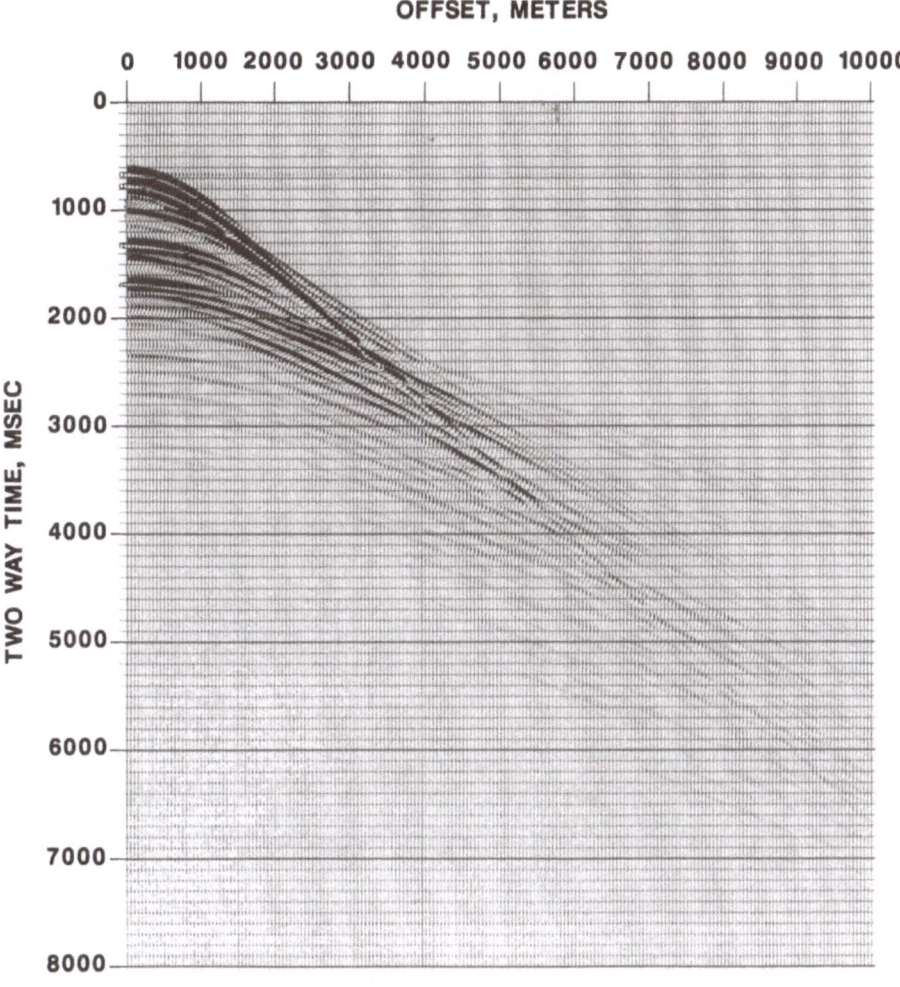

Fig. 2. Cartesian slant stack of the model shown in Figure 1. This seismogram corresponds to excitation by an infinite line source perpendicular to the plane of the problem.

ever, there are subtle amplitude and phase differences between these two seismograms, particularly at close distances from the sources. This, of course, is where the Bessel function in Equation (33) becomes substantially different from its asymptotic form. Figure 4 shows a detailed comparison between the Cartesian and cylindrical geometry models for two offsets.

To further illustrate the cylindrical slant-stack method, we present results from some wide-aperture common-midpoint (CMP) data shot by Gulf Research and Development Company over the West Florida Shelf. The experimental configuration is shown in Figure 5. Shooting constant-offset data with a ship separation of 8 km produced, after sorting, the CMP gather shown in Figure 6. These data have double coverage from 0–4 km and single coverage from 4–12 km. Because the

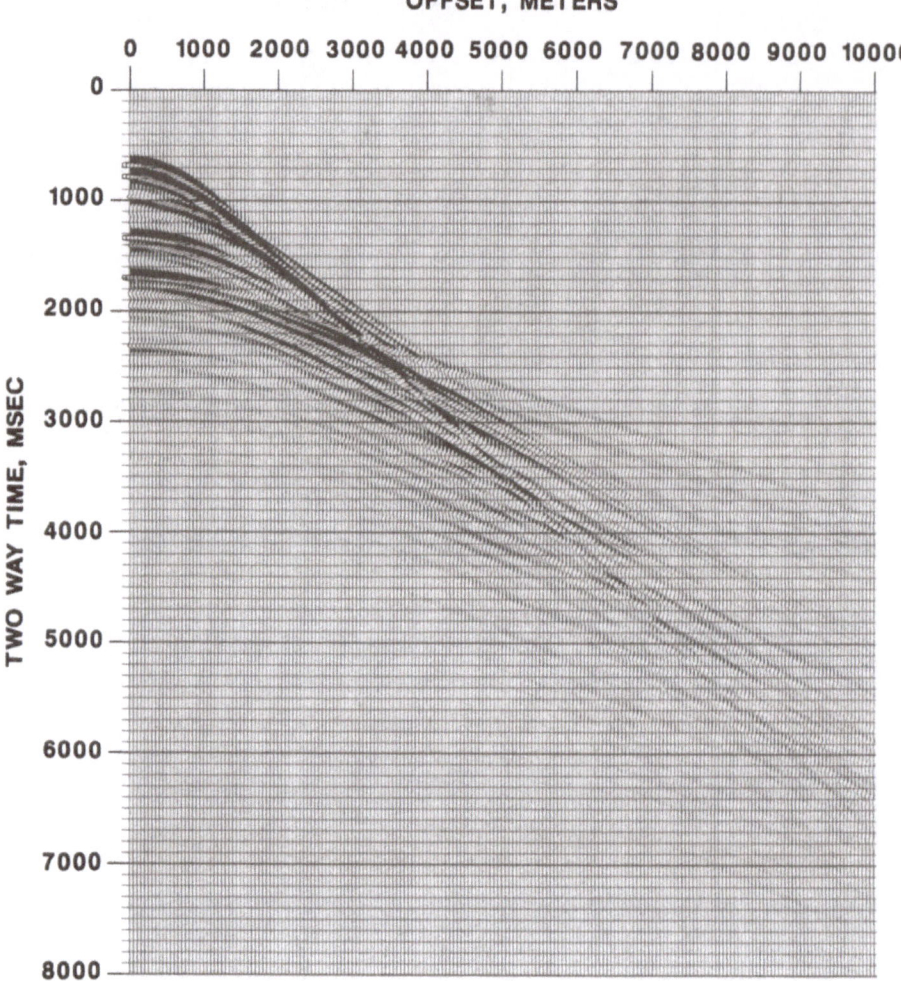

OFFSET, METERS

Fig. 3.   Cylindrical slant stack of the model shown in Figure 1. This seismogram corresponds to excitation by a point source at the origin. The plane of the problem is one slice of a cylindrically symmetric wave field.

ships were actually slightly closer than 8 km, there is a small region of triple coverage from 3.6–4 km.

To make the offset coverage more uniform, the CMP data were distributed into 50 m offset bins and summed. This section is shown in Figure 7. It is apparent that the 'binning' procedure significantly reduced the noise level in the 0–4 km range, allowing primary events to be seen at 0.8, 1.5, and 2.4 s.

A Cartesian slant stack of the data in Figure 7 is shown in Figure 8; the same slant stack after deconvolution is shown in Figure 9. In both figures there is severe edge-effect noise in the 0–0.5 s region. The slope of these linear features is 180 m, the minimum offset in the CMP gather. It is obvious from Figure 9 that predictive deconvolution cannot remove this noise. Stronger methods such as

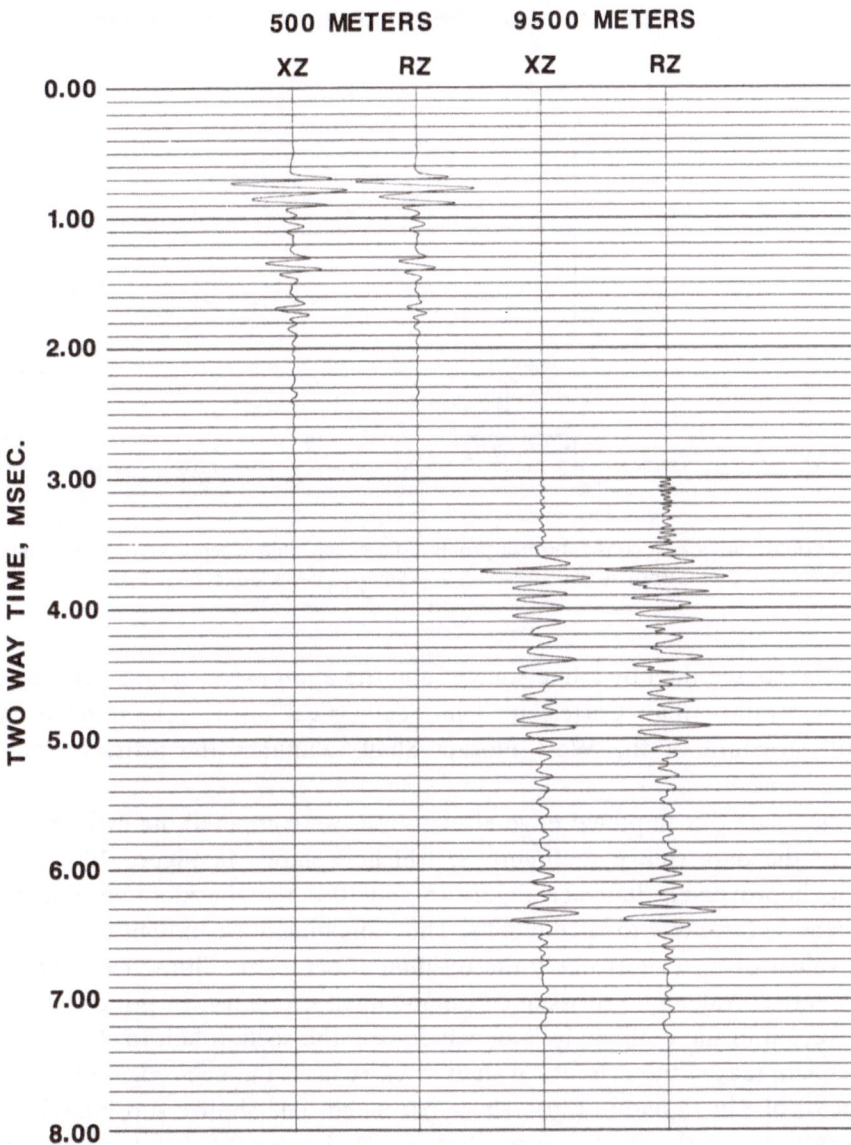

Fig. 4. Detailed comparison of two offsets from the Cartesian and cylindrical slant stacks of the model. First two traces at 500 m offset, second two traces at 9500 m offset. First and third traces are plane geometry, second and fourth are cylindrically symmetric spherical geometry. Corresponding events have slightly different amplitude and phase characteristics and different amplitude drop-off rates.

semblance weighting or offset windowing have traditionally been used to improve the signal-to-noise (S/N) ratio of Cartesian slant stacks.

The problem in this example, however, is neither spatial aliasing in the stacks nor edge effects, which are only symptoms. Instead, there is a fundamental error in the geometrical analysis. In fact, small air-gun arrays are better approximated

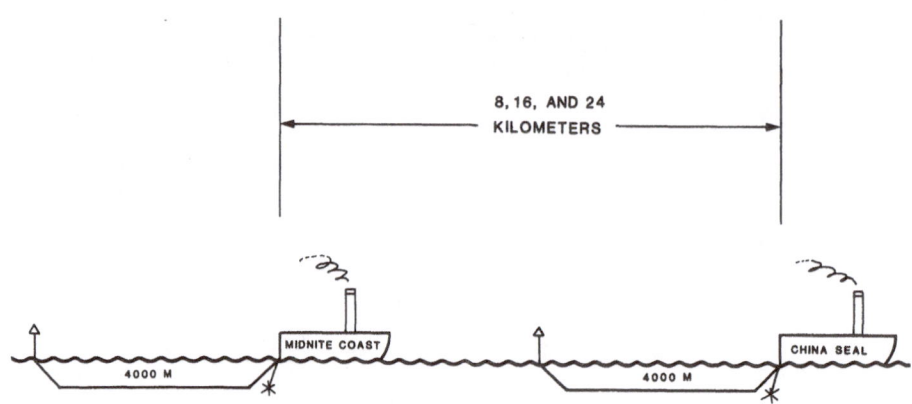

Fig. 5. Configuration of a constant-offset profile experiment. Ship separations of 8, 16, and 24 km extend the single-ship offset coverage to 4–12, 12–20, and 20–28 km, respectively.

by a point source than by a line source, and the appropriate slant stack should be cylindrical rather than Cartesian. This geometrical error, compounded by the difficult structure of the West Florida Shelf, amplifies the artifacts shown in Figures 8 and 9.

To eliminate the amplified edge-effect noise, we computed the cylindrical slant stack of the data shown in Figure 7, which is shown in Figure 10. The most obvious improvement between Figures 8 and 10 is the considerable reduction of the linear features due to edge effects. The remaining noise in Figure 10, visible at small values of p, is most likely the residual effect of the difference between the finite source array and our assumed point-source geometry. In principle, this can be corrected in the data analysis by windowing the section before slant stacking.

The main reason the cylindrical slant stack reduces the near edge-effect noise is the factor of r in Equation (30). These particular data show a substantial dropoff with offset due to the shallow water bottom. Weighting the data by r, as required in the cylindrical slant-stack method, reduces the contribution of the near offsets and, consequently, their truncation error. On the other hand, r-weighting amplifies the effect of data truncation at the far offsets, particularly with deep-water data. However, the latter problem can be solved by windowing data at large offsets which have poor S/N ratio anyway. Brysk and McCowan (1986) discuss the effects of windowing on cylindrical slant stacks and suggest an effective taper function. The cylindrical slant stack computed from our 12 km offset data exhibits little or no effect of the large-offset truncation problem.

Figure 11 shows the data of Figure 10 deconvolved. Since the multiples in a slant stack are truly periodic, deconvolution (as noted by Treitel et al., 1982) obviously enhances S/N ratio. The elliptical features in Figure 11, when compared

OFFSET, METERS

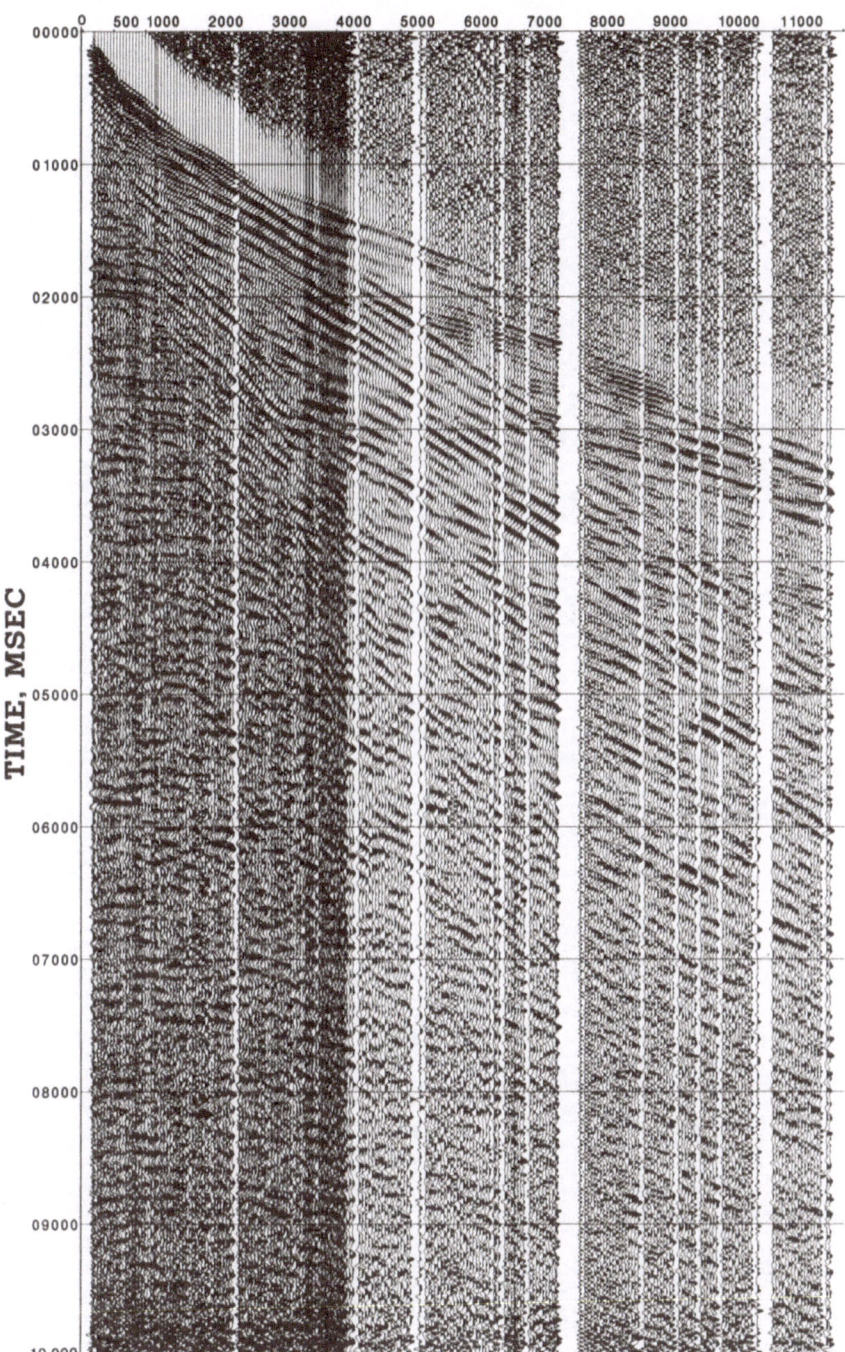

Fig. 6. A 12 km aperture CDP from the West Florida Shelf shot by Gulf Research and Development Corporation in 1984 using 2-ship techniques. Gaps and multiplicities in the offset coverage are caused by navigational practicalities.

OFFSET, METERS

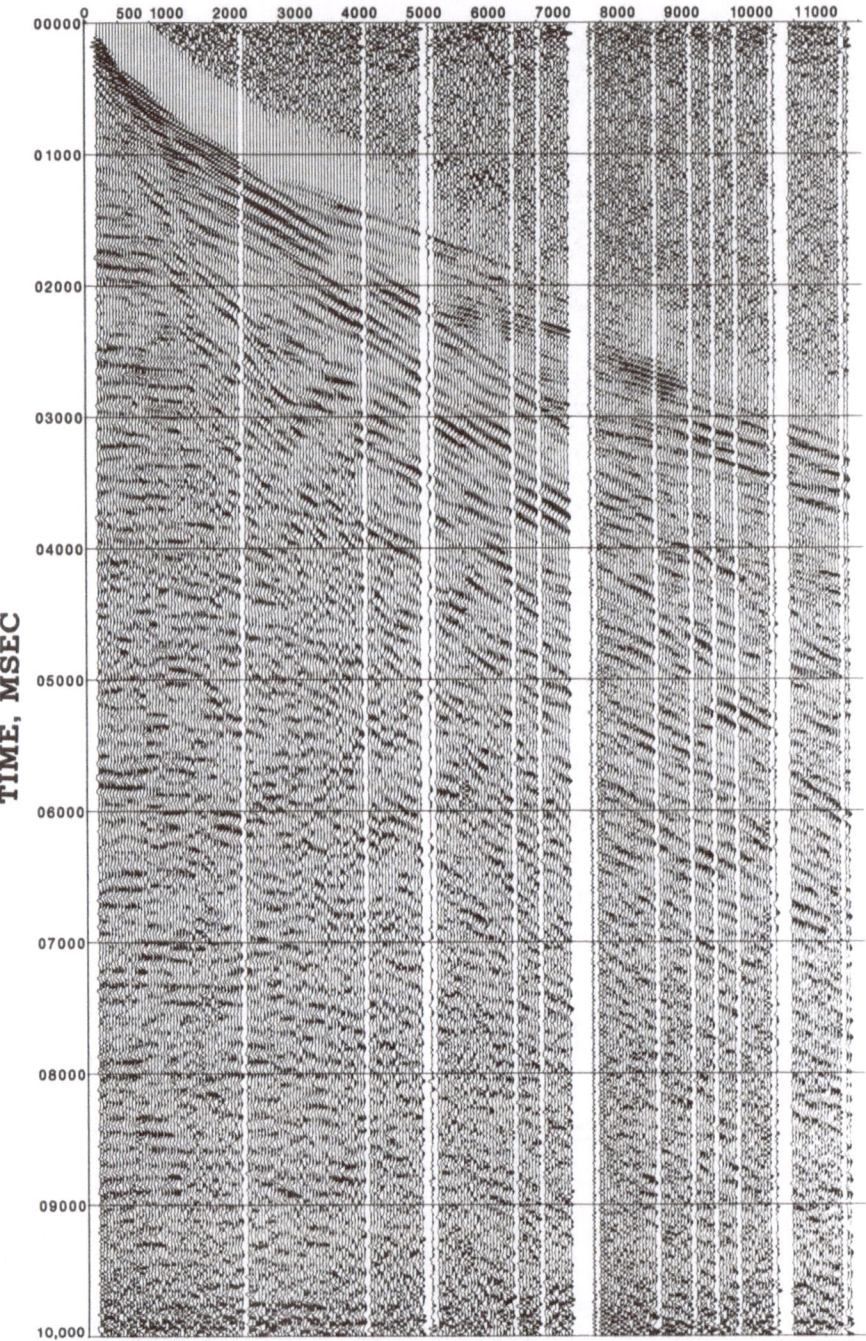

Fig. 7. The same CDP as in Figure 6 put on an even offset grid by 'binning' the data into 50 m bins. Each trace within an offset bin is corrected by NMO to the center of the bin and then summed and the result normalized by the number of traces in the sum.

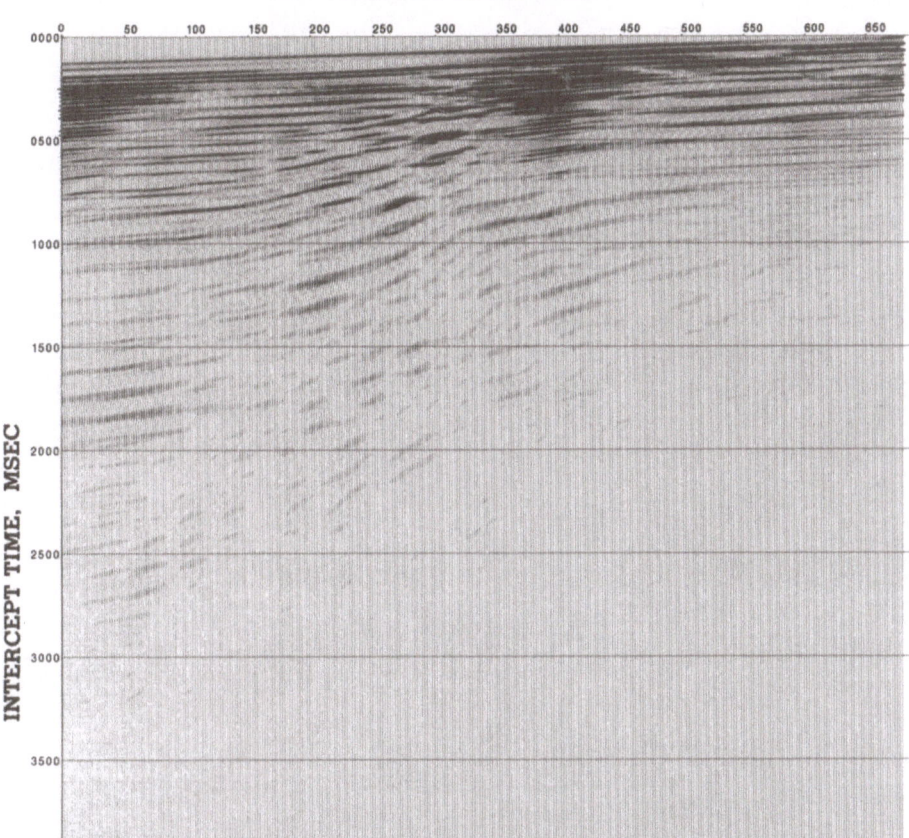

RAY PARAMETER, MSEC/KM

INTERCEPT TIME, MSEC

Fig. 8. An $X-Z$ geometry (Cartesian) slant stack of the data shown in Figure 7. The linear features between 0 and 0.5 s are caused by edge effects. These artifacts have slopes of 180 m, the minimum offset in the streamer cable.

to those in Figure 9, are noticeably sharper and simpler. They are, therefore, much more suitable for advanced analytical techniques such as amplitude analysis or inversion for earth structure.

Conclusions

The innovation introduced here is the procedure for calculating the cylindrical

RAY PARAMETER, MSEC/KM

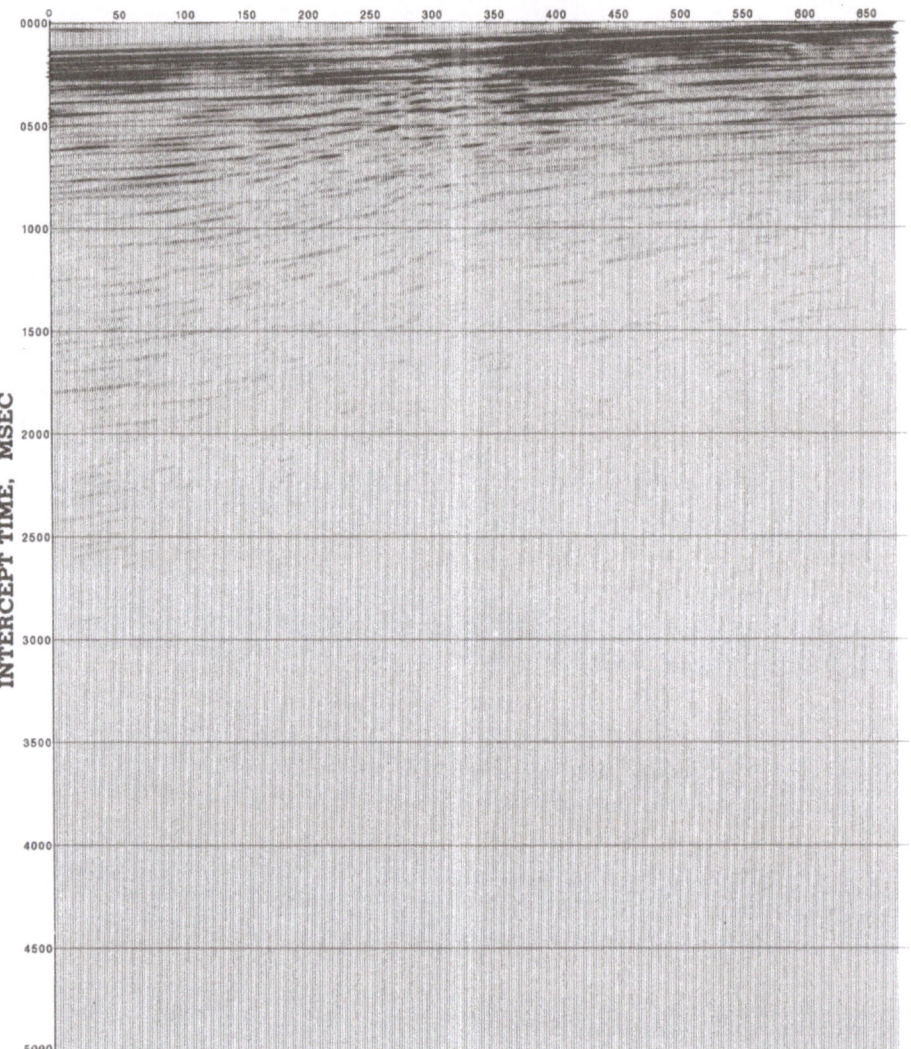

Fig. 9. The slant stack shown in Figure 8 after predictive deconvolution. Most of the edge effect artifacts remain visible.

slant stack from the common (line-source) slant stack. We chose this approach because it is computationally efficient and because it can be conveniently inserted into a preexisting program. It is a more practical alternative to the standard Bessel transform; the computational complications of which, especially in respect to aliasing, were extensively documented by Harding (1985).

The $\tau-p$ representation is widely recognized as a plane-wave decomposition, and we demonstrated this formally. As an important consequence, we proved that the plane-wave reflection and transmission coefficients apply exactly in the $\tau-p$

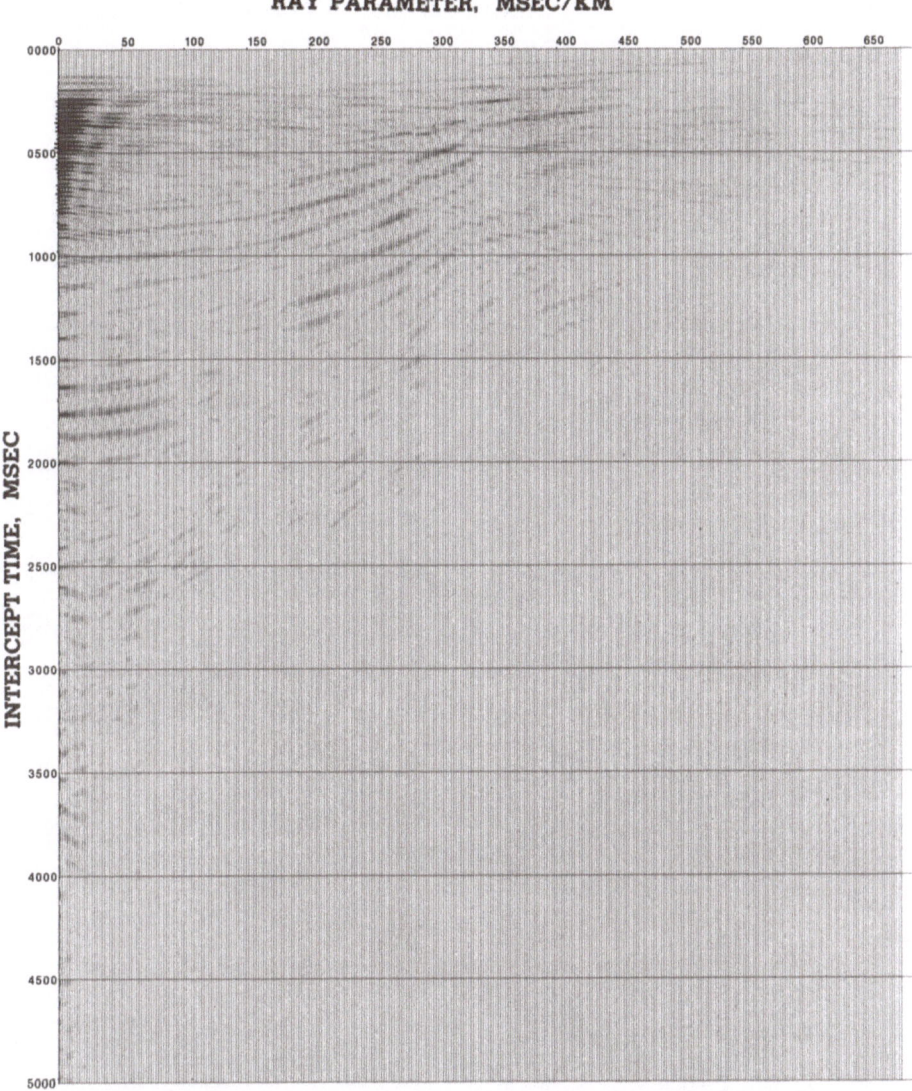

Fig. 10. An $R-Z$ geometry (cylindrical) slant stack of the data shown in Figure 7. The linear edge effect artifacts apparent in Figure 8 have disappeared.

domain, and we derived the associated phase relations. We thus introduced amplitude analysis in the $\tau-p$ domain with unambiguous reflectivities. There are discrepancies in the reflections of spherical as opposed to plane waves. For the conventional time display, the wave-curvature correction can sometimes alter the predicted amplitude-versus-offset variation. This may be sufficient to disrupt a bright-spot interpretation (Krail and Brysk, 1983). Rather than fine-tuning the reflection coefficients with geometry-dependent adjustments, it may well be simpler and safer to avoid this complication by transferring the analysis to the $\tau-p$ domain where all waves are planar.

RAY PARAMETER, MSEC/KM

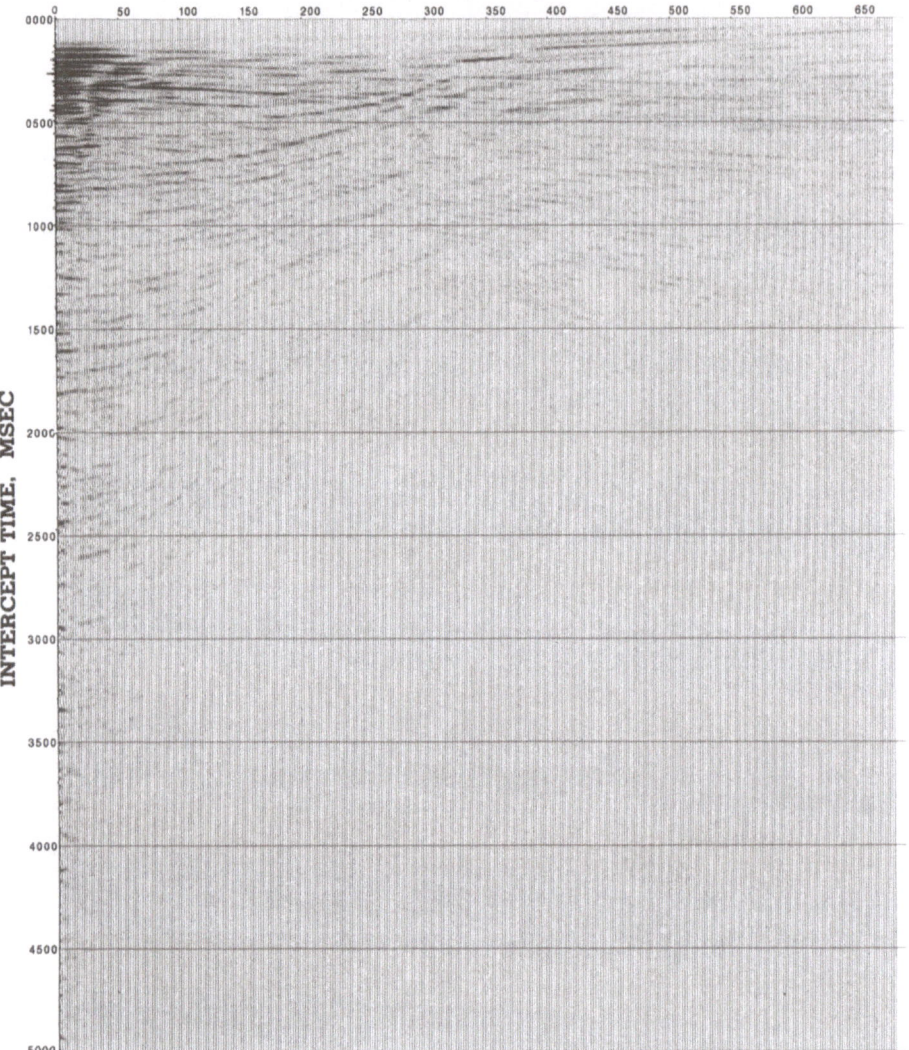

Fig. 11. The $R-Z$ slant stack in Figure 10 after predictive deconvolution. The elliptical features associated with plane parallel layers are now visible.

The actual data examples were obtained from Gulf's two-ship experiment (Figure 5) which provide high-quality data with an exceptionally wide aperture. Figures 8–11 display the results of basic true-amplitude processing, without cosmetics. They are valuable because they illustrate all the pertinent features with real data, whereas earlier discussions have had to depend largely on synthetics. There is no problem with far-offset truncation because the aperture exceeds the critical angle by a big margin. The near-offset edge effects are quite evident in the line-source $\tau-p$ processing, but they are largely mitigated in the point-source version because the geometrical factors act as a smoothing taper. There is a clear improvement in resolution with the cylindrical slant stack.

The point-source τ–p domain is useful for many kinds of reflection data analysis. Multiples can be effectively removed by predictive deconvolution because of their periodic nature as emphasized by Treitel *et al.* (1982). The direct inverse problem for one-dimensional structure is particularly simple when expressed in τ–p coordinates (Brysk and McCowan, 1986a). Finally, we note that the cylindrical slant stack properly accounts for the amplitude and phase effects of spherical spreading and thus should be useful as a spherical divergence correction.

Appendix A: Computation of the Cylindrical Slant Stack from Line-Source Slant Stacks

We describe the evaluation of the integral

$$S(p) = \int_0^p dp'\, R(p')(p^2 - p'^2)^{-1/2}, \tag{A.1}$$

when R is known at a uniform discrete set of values $p' = p_k = k\Delta p$, where k is an integer and Δp is a constant increment. This integral occurs in Equations (29) and (30). R is a line-source slant stack, obtained from traces to which a linear offset gain has been applied. S is the cylindrical slant stack.

The problem for numerical computation of the integral is the integrable singularity at the upper end. The singularity could be avoided by a change of variables, reverting to Equation (26). Equation (26) calls for R at the values $p' = p \cos \phi_i$ on a uniform grid of ϕ_i values, so R must be resampled onto a nonuniform p' grid. Furthermore, a different p' grid is required for each p, leading to a new resampling of R, which is not an appealing prospect.

Break up the integral into segments of length Δp. Within each segment, interpolate R as a function of p' and perform the integration analytically. Explicitly, consider $p = p_n = n\Delta p$. Within the segment $p_{k-1} \leqslant p' \leqslant p_k$, $k \leqslant n$, R is expressed by linear interpolation as

$$R(p') = kR_{k-1} - (k-1)R_k - (R_k - R_{k-1})p'/\Delta p, \tag{A.2}$$

with R_k denoting $R(p_k)$. The required analytic integrals are

$$\int_{p_{k-1}}^{p_k} dp'\, (p_n^2 - p'^2)^{-1/2}$$

$$= \arcsin \frac{k}{n} - \arcsin \frac{k-1}{n}, \tag{A.3}$$

and

$$\frac{1}{\Delta p} \int_{p_{k-1}}^{p_k} p'\, dp'\, (p_n^2 - p'^2)^{-1/2}$$

$$= [n^2 - (k-1)^2]^{1/2} - (n^2 - k^2)^{1/2}. \tag{A.4}$$

Adding the contributions of all the segments and writing S_n for $S(p_n)$, we find that

$$S_n = \sum_{k=0}^{n} C_{nk} R_k, \tag{A.5}$$

with the coefficients

$$C_{n0} = (n^2-1)^{1/2} - n + \arcsin \frac{1}{n}, \tag{A.6}$$

$$C_{nn} = (2n-1)^{1/2} - (n-1) \arccos \frac{n-1}{n}, \tag{A.7}$$

and, for $1 \le k \le n - 1$,

$$C_{nk} = [n^2-(k+1)^2]^{1/2} - 2(n^2-k^2)^{1/2} + [n^2-(k-1)^2]^{1/2} +$$
$$+ (k+1) \arcsin \frac{k+1}{n} - 2k \arcsin \frac{k}{n} + \tag{A.8}$$
$$+ (k-1) \arcsin \frac{k-1}{n}.$$

In particular, for $n = 1$,

$$C_{11} = 1 \quad \text{and} \quad C_{10} = (\pi/2) - 1. \tag{A.9}$$

Quadratic interpolation for R within each segment is intrinsically no more difficult. For convenience, the interpolation is specified so that Equation (A.5) is retained (i.e., no data points external to the range are used). For $n = 1$, this necessitates use of Equation (A.9). The derivation of the quadratic coefficients is considerably lengthier than the derivation of the linear coefficients, and it is omitted. The results for $n = 2$ are

$$C_{22} = -1+\pi/2, \qquad C_{21} = 4-\pi, \quad \text{and} \quad C_{20} = -3+\pi. \tag{A.10}$$

for $n > 2$, the end coefficients are

$$C_{n0} = \frac{1}{4} (n^2+4) \arcsin \frac{2}{n} + (n^2-4)^{1/2} - \frac{3}{2} n,$$

$$C_{n1} = \frac{1}{4} (n^2+12) \arcsin \frac{3}{n} - \frac{3}{4} (n^2+4) \arcsin \frac{2}{n} + \tag{A.11}$$

$$+ \frac{7}{4} (n^2-9)^{1/2} - 3(n^2-4)^{1/2} + 2n, \tag{A.12}$$

and

$$C_{nn} = \frac{1}{4} (3n^2-6n+4) \arccos \frac{n-1}{n} - \frac{1}{4} (3n-5)(2n-1)^{1/2}. \tag{A.13}$$

The next coefficients for $n > 3$ are

$$C_{n2} = \frac{1}{4}(n^2+24)\arcsin\frac{4}{n} - \frac{3}{4}(n^2+12)\arcsin\frac{3}{n} +$$

$$+ \frac{3}{4}(n^2+4)\arcsin\frac{2}{n} + \frac{5}{2}(n^2-16)^{1/2} -$$

$$- \frac{21}{4}(n^2-9)^{1/2} + 3(n^2-4)^{1/2} - \frac{1}{2}n, \qquad (A.14)$$

and

$$C_{nn-1} = \frac{3}{4}(3n^2-6n+4)\arcsin\frac{n-1}{n} -$$

$$- \frac{1}{4}(3n^2-10n+12)\arcsin\frac{n-2}{n} +$$

$$+ \frac{3}{4}(3n-5)(2n-1)^{1/2} -$$

$$- \frac{1}{2}(3n-8)(n-1)^{1/2} - \frac{\pi}{4}n(3n-4), \qquad (A.15)$$

while the corresponding value for $n = 3$ is

$$C_{32} = \frac{39}{2}\arcsin\frac{2}{3} + 3\sqrt{5} - \frac{15}{4}\pi - \frac{3}{2}. \qquad (A.16)$$

For $n > 4$, there are additional coefficients, with k satisfying $3 \leqslant k \leqslant n - 2$. They are

$$C_{nk} = \frac{1}{4}[n^2+2(k+1)(k+2)]\arcsin\frac{k+2}{n} -$$

$$- \frac{3}{4}[n^2+2k(k+1)]\arcsin\frac{k+1}{n} +$$

$$+ \frac{3}{4}[n^2+2k(k-1)]\arcsin\frac{k}{n} -$$

$$- \frac{1}{4}[n^2+2(k-1)(k-2)]\arcsin\frac{k-1}{n} +$$

$$+ \frac{1}{4}(3k+4)[n^2-(k+2)^2]^{1/2} - \frac{3}{4}(3k+1)[n^2-(k+1)^2]^{1/2} + \qquad (A.17)$$

$$+ \frac{3}{4}(3k-2)(n^2-k^2)^{1/2} - \frac{1}{4}(3k-5)[n^2-(k-1)^2]^{1/2}.$$

Note that the C_{nk} are independent of the data and can be precomputed.

Appendix B: Computer Program to Compute the Cylindrical Slant Stack
Coefficients

```
      SUBROUTINE SLCOEF(COEF,N)
      DIMENSION D(1000),E(1000),F(1000),COEF(N)
      DATA TWOPI/6.2831853/
C
C     CYLINDRICAL SLANT-STACK COEFFICIENTS (X4)
C     WITH QUADRATIC INTERPOLATION
C
C     WRITTEN BY HENRY BRYSK 10/84
C
      IF(N.GT.1000) STOP 1000
      NM=N-1
      GO TO (70,80,90), N
      PN=NM
      NM2=N-2
      PNM=NM2
      E(1)=-2.*PN
      DO 10 K=2,NM
      PK=K-1
      PKM=2.*(PK-1.)
      ROOT=SQRT(PN*PN-PK*PK)
      E(K)=(PK+PKM)*ROOT+(PN*PN+PK*PKM)*ATAN2(PK,ROOT)
   10 CONTINUE
      E(N)=TWOPI*PN*(.75*PN-.5)
      DO 20 K=3,NM
      D(K)=E(K+1)-E(K)
      F(K)=D(K)-D(K-1)
   20 CONTINUE
      COEF(1)=E(3)+3.*E(1)
      COEF(2)=D(3)-2.*(COEF(1)-E(1))
      COEF(N)=D(NM)-PNM*TWOPI
      IF(N-5) 60,50,30
   30 DO 40 K=4,NM2
      COEF(K)=F(K+1)-F(K)
   40 CONTINUE
   50 COEF(3)=E(5)-3.*D(3)+E(1)
      COEF(NM)=-F(NM)-D(NM)+PN*TWOPI
      RETURN
   60 COEF(3)=E(3)-2.*D(3)+E(1)+3.*TWOPI
      RETURN
   70 COEF(1)=TWOPI
      RETURN
   80 COEF(1)=TWOPI-4
      COEF(2)=4.
      RETURN
```

```
90  COEF(1)=2.*TWOPI-12.
    COEF(2)=4.-COEF(1)
    COEF(3)=TWOPI-4.
    RETURN
    END
```

References

Aki, K. and Richards, P. G., 1980, *Quantitative Seismology. Theory and methods*, W. H. Freeman.

Bessonova, E. N., Fishman, V. M., Ryaboyi, V. Z., and Sitnikova, G., A., 1974, The Tau method for inversion of travel times — I. Deep seismic sounding data, *Geophys. J. Roy. Astr. Soc.* **36**, 377–398.

Brekhovskikh, L., 1960, *Waves in Layered Media*, Academic Press.

Brysk, H. and McCowan, D. W., 1986a, Direct inversion of slant-stacked seismic data — Part I: Synthetic seismogram results, *Bull. Seism. Soc. Am.* **76**, 815–835.

Brysk, H. and McCowan, D. W., 1986b, Edge effects in cylindrical slant stacks, *Geophys. J. Roy. Astr. Soc.* **87**, 801–813.

Chapman, C. H., 1978, A new method for computing synthetic seismograms, *Geophys. J. Roy. Astr. Soc.* **54**, 481–518.

Chapman, C. H., 1979, On impulsive wave propagation in a spherically symmetric model, *Geophys. J. Roy. Astr. Soc.* **58**, 229–234.

Chapman, C. H., 1981, Generalized Radon transforms and slant stacks, *Geophys. J. Roy. Astr. Soc.* **66**, 445–453.

Deans, S. R., 1983, *The Radon Transform and Some of Its Applications*, John Wiley, New York.

Diebold, J. B. and Stoffa, P. L., 1981, The traveltime equation, tau-p mapping, and inversion of common-midpoint data, *Geophysics* **46**, 238–254.

Fryer, G. J., 1980, A slowness approach to the reflectivity method of seismogram synthesis, *Geophys. J. Roy. Astr. Soc.* **63**, 747–758.

Harding, A. J., 1985, Slowness-time mapping of near-offset seismic reflection data, *Geophys. J. Roy. Astr. Soc.* **80**, 463–492.

Henry, M., Orcutt, J. A., and Parker, R. L., 1980, A new method for slant stacking refraction data, *Geophys. Res. Lett.* **7**, 1073–1076.

Krail, P. M. and Brysk, H., 1983, Reflection of spherical seismic waves in elastic layered media, *Geophysics* **48**, 655–664.

Muller, G., 1971, Direct inversion of seismic observations — Z. *Geophys.* **37**, 225–235.

Phinney, R. A., Chowdhury, K. R., and Frazer, L. N., 1981, Transformation and analysis of record sections, *J. Geophys. Res.* **86**, 359–377.

Schultz, P. S. and Claerbout, J. F., 1978, Velocity estimation and downward continuation by wavefront synthesis, *Geophysics* **43**, 691–714.

Stoffa, P. L., Buhl, P., Diebold, J. B., and Wenzel, F., 1981, Direct mapping of seismic data to the domain of intercept time and ray parameter — A plane-wave decomposition, *Geophysics* **46**, 255–267.

Treitel, S., Gutowski, P. R., and Wagner, D. W., 1982, Plane-wave decomposition of seismograms, *Geophysics* **47**, 1375–1401.

Douglas W. McCowan,
Mobil Research and Development
* Corporation,*
Dallas Research Laboratory,
13777 Midway Road,
Dallas TX 75344-4312,
U.S.A.

Henry Brysk,
CGG,
2500 Wilcrest,
Suite 200,
Houston, TX 77042,
U.S.A.

ROBERT H. TATHAM

Tau–p Filtering

Abstract

Examination, evaluation and processing of seismic data in the two-dimensional τ–p domain offers many advantages over analysis and processing in the originally recorded X–t domain or other transformed domains. Since the slowness p (1) is the reciprocal the horizontal phase velocity, (2) represents the apparent angle of emergence at the surface, and (3) in a flat-layered earth it is also a constant conserved quantity (the ray parameter) along the raypath, we can readily address (1) antenna and array problems, (2) angle of incidence effects, and (3) problems (such as multiples) associated with the ray geometry. Examples of these applications include attenuation of ground roll, separation of P and S-waves where the angle of incidence defines the efficiency of mode conversion, and more efficient attenuation of multiples where they are exactly periodic in the τ–p domain.

The present paper reviews and discusses some of these applications, with real data examples of each. It also discusses aspects of the τ–p transform relevant for such applications and an alteration of the forward transform that improves its filtering characteristics.

Introduction

One of the primary purposes of transforming two-dimensional seismic records to the τ–p domain is to separate coherent events in the transformed domain that normally interfere with one another in the originally recorded domain. This separation allows for effective isolation, and subsequent filtering, in the two-dimensional domain. Further, other physical properties, such as angle of incidence effects and periodicity of multiples, may be readily exploited in the τ–p domain. The thrust of the current discussion is entirely on such filtering in the τ–p domain. To describe the fundamental reasons for the separation of normally interfering events in the transformed space, we will devote some discussion to descriptions of the transform process, using a slant-stack implementation and a flat layered earth model. This discussion should provide some insight into the nature of the seismic events in question, and the effect of their filtering. The flat-layered earth assumption (a 1D earth) does not present serious limitations to the discussion, since (1) for exploration purposes, the earth is nearly flat, with sub-parallel layers, (2) after a CMP gather, the data approach a 1D model, and (3) for

Paul L. Stoffa (ed.), Tau-p: A Plane Wave Approach to the Analysis of Seismic Data, 35—70.
© 1989 *by Kluwer Academic Publishers.*

many problems to be addressed, such as ground roll and near-surface scattering, assumptions about layer geometry really don't matter anyway.

One of the quantities typically applied in the description of two-dimensional filtering (describing a filter in terms of reflection time and apparent velocity) is the apparent horizontal phase velocity. In fact, frequency filtering in one dimension extends to $f-k$ filtering in two-dimensions, and the band-limits of the $f-k$ filter are usually expressed in terms of apparent horizontal phase velocities. In $\tau-p$ filtering, velocity (or equivalently the ray parameter p) is used to describe the filter limits, but this can be applied in a time-varying sense. That is, zero-offset arrival time τ, in conjunction with p, is also used to describe filter limits.

Besides the advantages of describing two-dimensional filters in a time-varying sense, $\tau-p$ filtering permits the describing filters in a physically satisfying way. Unlike $f-k$ filtering, where each sample in the transformed space is complex (with real and imaginary components to contend with), the $\tau-p$ domain is entirely real, and time traces contain familiar wavelets and continuous seismic events. Thus, examination and interpretation of data in the transformed space is comfortable for most practicing geophysicists, or should be once they get used to visualizing p, which is slowness or reciprocal velocity.

Another important aspect of the $\tau-p$ domain is that the ray parameter p which, for constant near-surface velocity (e.g., the water layer in a marine setting), represents a single angle of incidence at the surface for the entire time trace. This is especially useful for addressing events, such as mode conversion and response to source and receiver arrays, which are strongly dependent upon angle of incidence. Also, for a given angle of incidence, multiples, at least from a particular multiple generator in a 1D earth, are exactly periodic. Thus, deconvolution and multiple attenuation schemes based on periodicity of multiples (e.g., predictive deconvolution) should work better on a constant p-trace than on the conventional constant offset trace.

General descriptions of the $\tau-p$ transform have been presented by many investigators and in several different contexts. Besides other authors in the present volume, these include Schultz and Claerbout (1978), Stoffa et al. (1981), Phinney (1981), Phinney et al. (1981), Alam and Austin (1981), Tatham et al. (1981, 1983b), and Treitel et al. (1982).

The objective of the present paper is to suggest filtering applications in the $\tau-p$ domain. Such applications do require some intuitive insight into the physical aspects of the transform and, thus, I will briefly review the transform itself in terms of a slant stack. Use of the slant stack description, rather than a more formal complete plane wave decomposition, allows for simple limitations in the data that are employed in the transform. Filtering may be thought of as a limitation in the data to either eliminate some undesired region of the data set or isolate some particularly useful subset of the data. Consideration of a full invertable transform is thus not expanded upon. Of course, if filtering limits are set infinitely large, the slant stack should be the same as the more complete treatments. This will lead to an understanding of how and why interfering events in one domain separate in the other, and just how the physical significance of the ray parameter allows effective filtering, particularly for the 1D earth assumption.

Description of the Transform in Terms of Slant Stack

APPARENT HORIZONTAL PHASE VELOCITY

Before proceeding with the slant-stack discussion, we require a few definitions, especially of velocity. Horizontal distance along the surface is given by x. Generally, at least for a shot record, the origin of x is taken as the shot location and, thus, x also represents the source-receiver offset. The seismic propagation velocity (equal to the hyperbolic normal moveout velocity for flat reflectors in a constant velocity medium) can be related directly to the apparent horizontal phase velocity by considering the angle of incidence of a ray to the recording surface. Figure 1 outlines the geometry of a plane wavefront intersecting a surface that contains the recording sensors. The ray direction, normal to the wavefront, is also shown. The angle of incidence (i) is expressed, for a flat layered earth, as the angle between the vertical and the ray direction. Since the wavefront is normal to the ray, i is also the angle between the wavefront and the horizontal. For a small distance dx, the ray travels a small distance V dt and

$$V\frac{dt}{dx} = \sin(i) \quad \text{or} \quad \frac{dt}{dx} = \frac{\sin(i)}{V}. \tag{1}$$

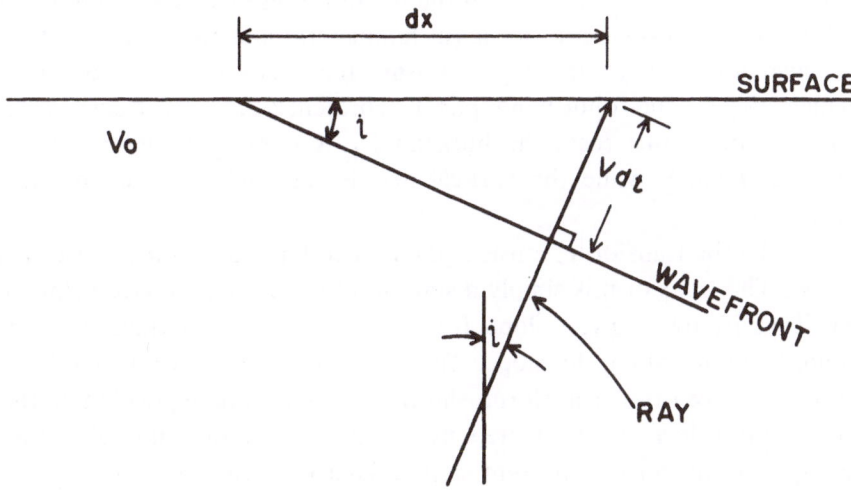

Fig. 1.   Plane wavefront on a seismic wave intersecting the surface of the earth. Angle i is (1) the angle between the vertical and the ray associated with the wavefront, and (2) the angle between the horizontal surface and the wavefront itself. V is the propagation velocity of the seismic wavefront, and dx/dt is the apparent horizontal phase velocity of the wave as it sweeps across the horizontal surface (after Tatham, 1984).

Thus, dx/dt = the apparent horizontal phase velocity $V/\sin(i)$, and we have the desired relation between horizontal phase velocity dx/dt and propagation velocity V.

RAY PARAMETER p

The quantity dt/dx in (1) is often referred to as the ray parameter p, so

$$p = \frac{dt}{dx} = \frac{\sin(i)}{V}.$$

This quantity is also referred to as the horizontal slowness, or reciprocal velocity. Slowness is a common term in antenna theory.

As an aside, Snell's law is often expressed as

$$\frac{\sin(i_1)}{V_1} = \frac{\sin(i_2)}{V_2} = \cdots = \frac{\sin(i_n)}{V_n}$$

for a ray tracing from medium 1 to medium n. Since $p = \sin(i)/V$ for all i's and V's, stating that p is constant for a given ray is just a restatement of Snell's law. For this reason, p is a popular parameter for describing rays in ray theory.

SLANT STACK

The construction of the $\tau-p$ transform by means of a slant stack is illustrated in Figure 2. Also described are the mapping of events from one two-dimensional domain to another. The left panel in the figure represents a conventional seismic record whose dimensions are offset (in kilometers) and two-way reflection time (in seconds) at that offset. In a typical seismic record, each time trace represents a particular offset. The right-hand panel represents the $\tau-p$ transform of the seismic record. In this case, the horizontal axis is ray parameter p (in microseconds per meter), while the vertical axis is zero offset reflection time τ (in seconds).

To describe the transform, consider the dashed lines, at a constant slope, in the left panel. The transform is simply a sum of all traces along a given slope (slant-stack). That is, for a given slope (or slant) $p = dt/dx$, a trace is formed by summing all points along the slope. This summed trace, shown to the left of the panel, is one trace of the transform, shown at its appropriate position on the right panel. The transform is completed by forming an entire suite of p traces by summing along numerous different slopes. That is, each time trace of the transform is simply the sum of all input traces taken along a particular slant.

Figure 2 also illustrates the correlations of typical seismic events in the two domains. In the left panel, representing the usual $X-t$ record space (either field records or CMP gathers), we see a ground-roll (direct) arrival D, two linear refractions H_1 and H_2, and three hyperbolic reflections. Note that the two linear

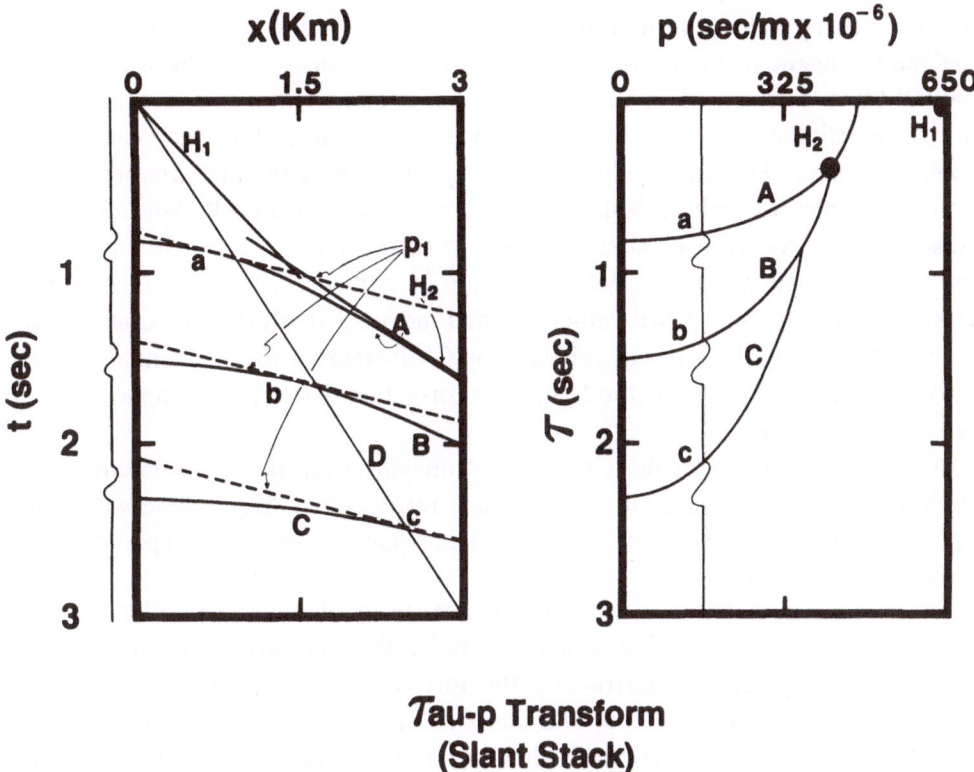

*T*au-p Transform
(Slant Stack)

Fig. 2. Construction of the slant-stack (τ–p transform). Left panel schematically represents events on a seismic field record. Hyperbolae *A*, *B*, and *C* represent reflections. Linear events *D*, H_1, and H_2 represent a direct arriving ground-roll (*D*) and two refractions. To construct the slant-stack transform, all traces in the seismic record are summed along a linear slope (slant-stack) to yield a single trace for the transform. The sample trace in the left panel, generated by summing along slant p_1, is a single trace in the right panel. The right panel, representing the equivalent τ–p space, has traces in units of time but each trace is a given slant, or slope, *p*. The quantity *p* has units for reciprocal velocity (sec/m). Events that interfere in the left panel are separated in the right panel. In particular, reflection hyperbolas *A*, *B*, and *C* become ellipses *A*, *B*, and *C*. The linear event *D* has a slope of 980 μsec/m and is thus beyond the range of *p* values in this transform. The refractions H_1 and H_2 become points (or small regions) H_1 and H_2 in τ–p space. Further, the arrival time τ is just the zero-offset intercept time on the refractions (after Tatham *et al.*, 1983b).

refraction events have constant slope d*t*/d*x* along their trajectories and a single zero-offset arrival time. Thus, they transform to single points, denoted H_1 and H_2 in the τ–p domain. The ground roll, also a linear event, does not appear in the right panel because its velocity is less than that represented by the maximum value of *p* in the transformed domain. Thus, the ground-roll is not included in this particular τ–p transform, and inversion from the τ–p domain back to the *X*–*t* domain will reconstruct all events originally observed except the ground-roll. This presents a simple method of filtering ground roll from seismic records.

The reflection hyperbolas, with differing slopes d*t*/d*x* along their trajectories, transform to ellipses in the τ–p domain. Note the tangent points *a*, *b*, and *c*, for the illustrated p-value, on the reflection hyperbolas. On the ellipses A, B, and C, the same tangent points correspond to the intersections of the ellipses with the plotted trace. The reflection ellipses do not cross one another in $\tau-p$ space, even if the hyperbolas do cross in record space. They may, however, meet at a common point. Thus, in transforming to $\tau-p$, we have separated crossing reflection hyperbolas as well as isolating refractions and ground-roll. Since Diebold (this volume) has given us nearly every $\tau-p$ travel time equation we could ever use, we can apply an exact NMO correction in $\tau-p$ space. If the $\tau-p$ transform was constructed from a CMP gather, normal moveout (NMO) corrections can be applied (i.e. along the ellipse), followed by CMP stacking to form a final stacked trace. This is an excellent method of NMO correction that avoids complications of crossing hyperbolas.

Another point to be made is the role of imaging from the $\tau-p$ domain. Keep in mind that each p, or angle of incidence, represents a single plane wavefront and, thus, wavefronts are separated, and imaging is, for many applications, a straightforward procedure.

Since the $\tau-p$ transform does have an inverse, as discussed elsewhere in this volume, filtering of interfering seismic events on the conventional record is easily accomplished by simply constructing the forward transform, removing the undesired events that are isolated, and constructing the inverse transform.

Before proceeding with some modifications of the transform for filtering purposes, we will review the definition and application of the forward transform, and examine a simple example of the $\tau-p$ transform, as well as applications to filtering. Figure 3 shows the original $X-t$ record of a walk-away analysis performed in a physical model experiment (Tatham et al., 1983a). Scaled offsets in the record range from 300 m to 10 km, in increments of about 30 m (100 ft). The entire record consists of 290 traces, each 3 sec long, sampled at a time interval scaled to 0.001 sec. The model was designed to observe mode-converted shear waves at large offsets, as well as conventional P-wave reflections, in a shallow-water marine environment. The letters $A-D$ identify four sets of reflections, known from the model, for both P-wave and S-wave propagation paths. The P-wave reflections are at the shorter offsets, near the left-hand side of the figure, while the S-wave reflections are seen at larger offsets. The labels $A-D$ identify the reflector, and the S-wave reflection from an interface occurs at a greater time than the P-wave reflection from the same interface. The model is deeply immersed in a water-filled tank, and marine water depth is scaled to 75 m. This is accomplished by situating the source and receiver a scaled 75 m above the solid model. In addition to the reflections, note the refraction (from the model water bottom) and the strong direct arrival.

For present considerations, note that the seismic events observed in Figure 3 include a linear event (energy arriving directly form the source) with velocity about 3650 m/sec, (the P-wave propagation velocity in water scales to 3650 m/sec) and a critical angle refraction (linear event above the direct arrival) with velocity

OFF-SET(Km)

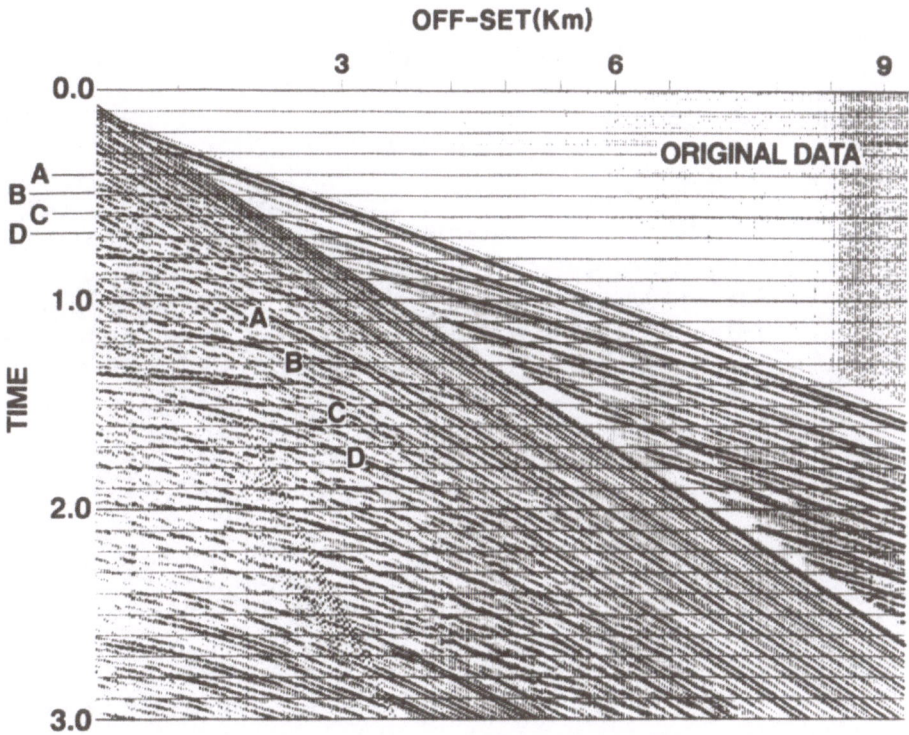

Fig. 3. Record of walk-away analysis of seismic events in a physical model of a marine experiment. Offsets scale from 300 m to 10 km in steps of about 30 m (100 ft). Water depth scales to 75 m. Note strong direct arrival, refraction and *P*-wave reflections at shorter ranges, *S*-wave (mode-converted) reflections at larger offsets and interference caused by the direct arriving event. The four reflecting interfaces are indicated, *A – D*. The *S*-wave reflections are only seen at large offsets and at about twice the reflection time as the *P*-wave reflections. The model was designed to study mode-converted shear-wave (*SV*) reflections (after Tatham *et al.*, 1983a).

6000 m/sec. *P*-wave reflection hyperbolas are observed on the near-offset traces, and reflection hyperbolas with velocities lower than those of *P*-waves (mode-converted shear waves) are observed at offsets from about 2500 to 4500 m. Beyond 4500 m, very little information is present in this 3 sec record. To some degree, this lack of deeper data will affect the geometric assumption of the $\tau-p$ transform for large offset data.

Figure 4 shows a $\tau-p$ transform (slant stack) applied to this original shot-oriented data set. The individual traces in the transform domain are in units of time (zero-offset time) and involve no stretch during the transform process. Hence, the wavelet character is preserved. Note how well all events which interfere with one another in Figure 3 are separated in Figure 4. Further, the $\tau-p$ domain, as we have defined it, is entirely real; there is no need to examine just the amplitude spectrum when both real and imaginary components need to be considered. As discussed earlier, for constant near-surface velocity (like the water

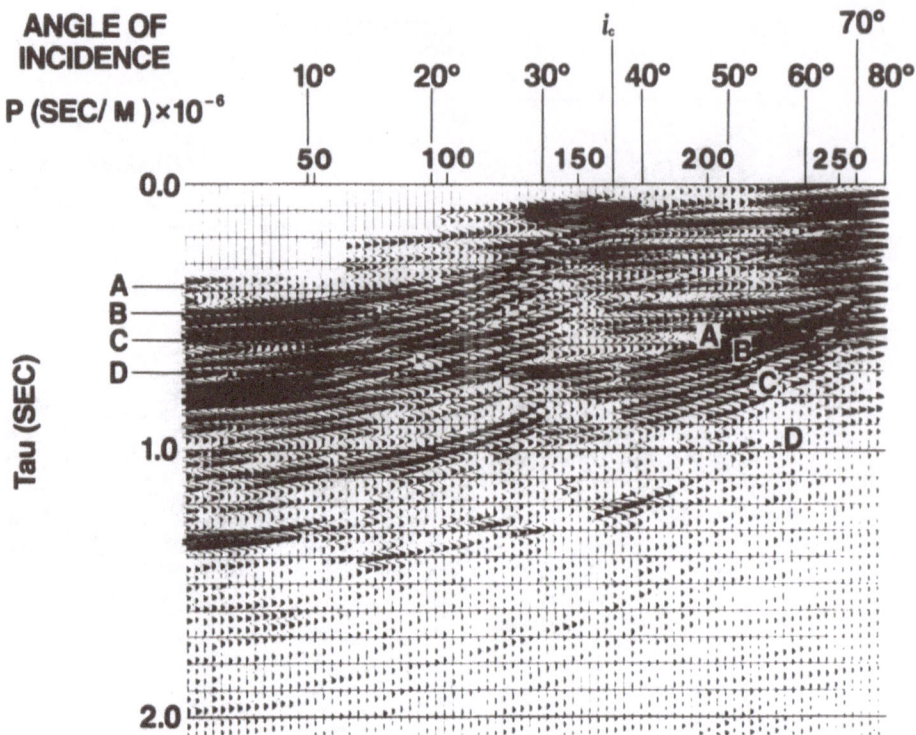

Fig. 4.   Tau-p transform (slant-stack) of the original shot-oriented field record shown in Figure 3. The
individual traces in the transform domain are in units of time (zero-offset time) and involve no stretch
during the transform process. Each trace represents a single ray parameter $p = \sin i/V$. If the near-
surface velocity is known, as in the marine case, each p represents a known angle of incidence. In this
example, the water velocity in the model scales to 3650 m/sec. Note that events which interfere with
one another in the record space are well separated in the τ–p domain, allowing their separation on a
inverse transform back to the record space. Reflection hyperbola in the record space transforms to
ellipses in the τ–p domain; refractions, direct arrivals and ground-roll transform to points or small
regions about a point (after Tatham *et al.*, 1983a).

in this example), each p value corresponds to a known angle of incidence. Thus,
events which are strongly dependent upon the angle of incidence, such as mode-
converted shear waves, can readily be isolated on that basis. Angles of incidence
(at the surface) for each p-value are shown at the top of the figure. The separation
in the τ–p domain of events that interfere with one another in x–t record space
allow filtering on the inverse transform back to x–t space. We see the reflection
ellipses corresponding to hyperbolas and the point-like concentration of energy
associated with the direct arrival (70–80°) and the critical angle refraction (35°).
S-wave reflections are especially apparent beyond the critical angle.

As an example of filtering on the basis of angle of incidence, consider the
results shown in Figure 5. This is the inverse transform of the data shown in
Figure 4, but limited to p-values associated with angles of incidence less than the
critical angle at the water bottom. This limitation greatly attenuates the direct

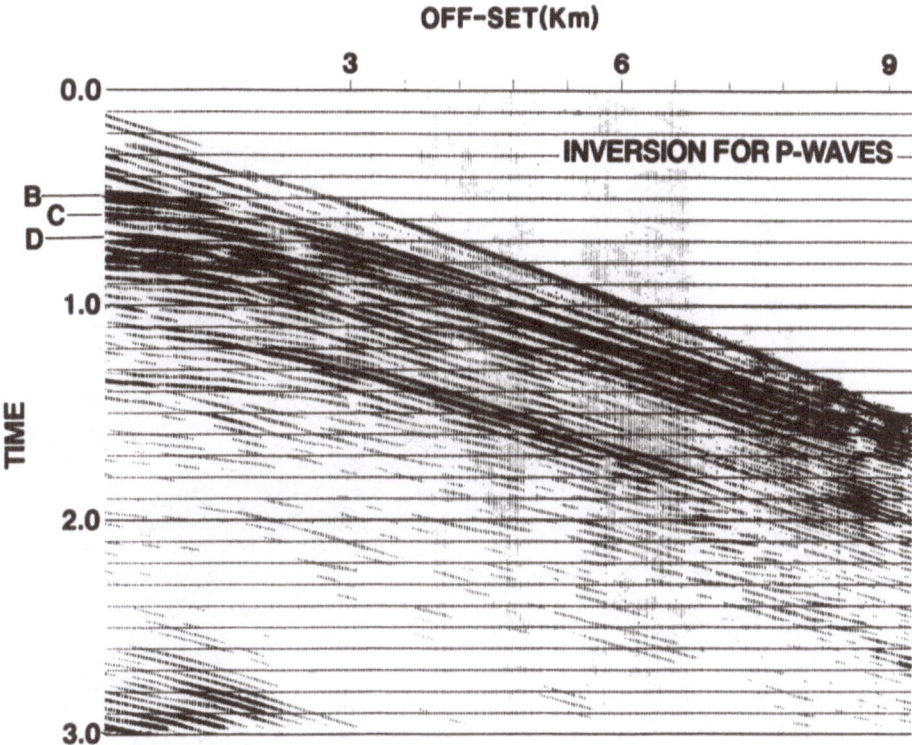

Fig. 5. Inverse transform for angles of incidence less than the critical angle isolates *P*-wave reflections, attenuating the direct arrival and refraction events (after Tatham *et al.*, 1983a).

arrival and refraction events. Since less than 50 *p* traces were used in the inverse transform from τ–*p* to this record, this reconstructed record can only be an approximation of the original data. This approximation, however, does emphasise the *P*-wave reflections and the attenuation of mode-converted shear-wave reflections.

Figure 6 shows a similar inverse transform, but for angles of incidence greater than the critical angle at the water bottom. Thus, no conventional *P*-wave energy is possible in the subsurface. In this case, the mode-converted shear wave reflections are well isolated. These two records show a simple application of separation of *P* and *S*-wave reflections in the marine environment.

At this point, we have separated *P* and *S*-wave reflections. Velocity inversions, for both *P* and *S*-wave velocity profiles are possible in the τ–*p* domain. Further, by considering interval travel times τ for both *P*- and *S*-waves at a single *p* value (i.e. a single ray), we can form a travel-time ratio Ts/Tp and, hence, have an immediate estimation of Poisson's ratio for that interval. With redundancy over several *p* values, we have several nearly independent estimates of Poisson's ratio and, thus, a measure of the reliability of the estimation.

For completeness, an inverse transform for all of the data shown is Figure 4 is

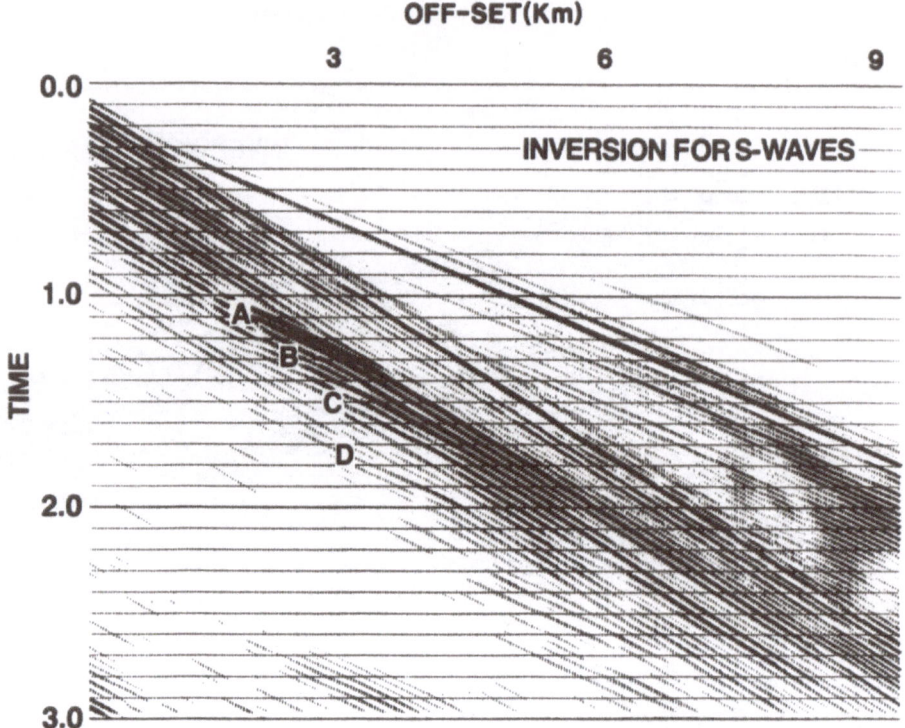

Fig. 6. Inverse transform for angles of incidence greater than the critical angle eliminates *P*-wave reflections, isolating shear-wave reflection hyperbola (after Tatham *et al.*, 1983a).

shown as Figure 7. This example demonstrates the robustness of the entire forward/inverse procedure. The most notable difference between the original record and this doubly transformed record is the diminished amplitude of the direct arrival. This is due to the point-like nature of the events in τ–p space, and p-sampling too sparse to maintain the dynamic range of the strong linear event. Furthermore, since 290 traces were reduced to fewer than 100 traces in the forward transform, the full dynamic range of the entire record was not preserved. The limit in dynamic range is especially apparent regarding later-arriving events. The character and position of the reflection events, however, are preserved. This particular example illustrates the need to maintain sufficient p-sampling.

Hyperbolic Velocity Filtering as Part of The Forward τ–p Transform

An additional set of filtering criteria can be applied to the forward τ–p transform by supplying some reasonable range of moveout velocities. This range of velocities, often known from regional geological knowledge, represents additional fundamental information and allows the restriction of slopes, or p-values, that a given data sample may contribute to during the transform process.

This modification to the forward transform is simply limiting the data, or

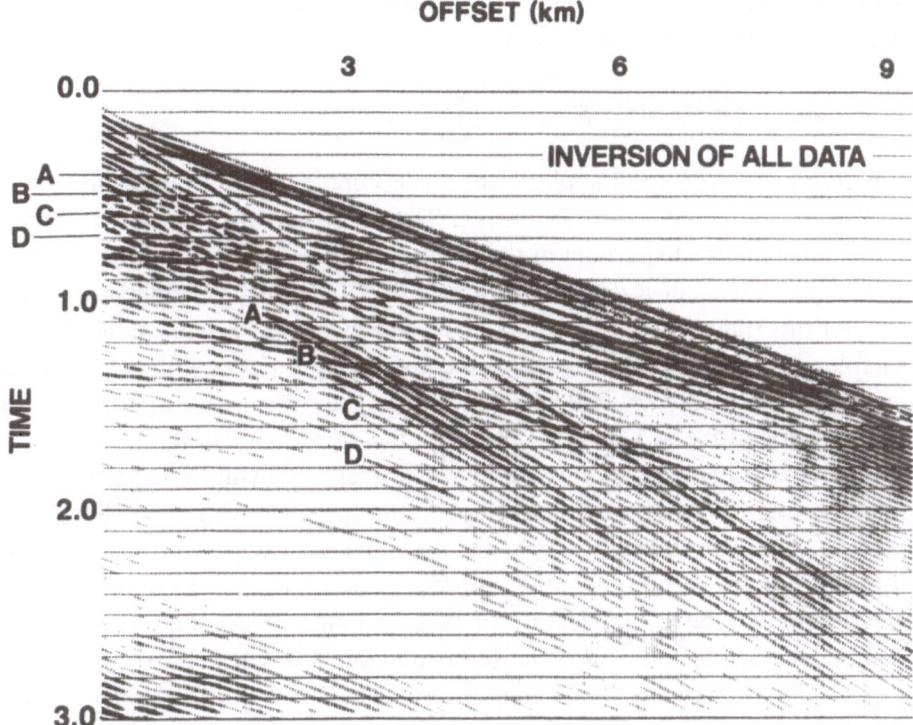

Fig. 7. Inverse transform of the entire data set. The most notable difference between the original record and this doubly transformed record is the diminished amplitude of the direct arrival. This is due to the point-like nature of the direct arrival in τ–p and insufficient p-sampling to maintain the dynamic range of the event (after Tatham *et al.*, 1983a).

reducing the range of the integration, in the forward transform. Rather than a change to finite limits in the integration, or limits defined by the aperture of the recording array, we apply limits based on anticipated ranges in regional velocity. This can also be thought of as a change in the kernel of the transform where a pair of step functions, or a box-car function, is included. The limits of the box-car functions vary with X, p, and t. Thus, this filtering option adds additional dimensions p, X and t, to the filtering process. Filtering by solely limiting the data in the τ–p domain ignores this additional and, as will be seen, very useful expansion.

To define these limits, which vary with X, t, and p, assume that all events of interest are reflections defined by the usual normal moveout hyperbola

$$t^2 = t_0^2 + \frac{x^2}{V^2}.$$

Differentiation yields

$$2t\,\mathrm{d}t = \frac{2X\,\mathrm{d}x}{V^2} \quad \text{or} \quad p = \frac{\mathrm{d}t}{\mathrm{d}x} = \frac{X}{tV^2}.$$

Thus, the upper and lower limits in velocity V_{max} and V_{min} define the lower and upper limits in appropriate p values

$$p_{min} = \frac{X}{tV_{max}^2} \quad \text{and} \quad p_{max} = \frac{X}{tV_{min}^2}.$$

Figure 8 illustrates the application of this concept. This example assumes V_{max} = 3750 m/sec and V_{min} = 2250 m/sec. The range of NMO velocity is 3000 m/sec ±25%. The flatter hyperbola in the figure is used as the upper velocity limit, with t_0 = 2 sec. Consider a data point at an offset of 3 km and reflection time 2.154 sec. At this sample, p_{min} = 97.4 μsec/m and p_{max} = 270 μsec/m. Thus, this data sample will only contribute to sums in constructing p-traces with values between p = 97 μsec/m and p = 270 μsec/m. Of course, the limits in p-values will be different for every t–X sample, even for the same velocity limits.

Applying the limits in p-values is effectively applying a hyperbolic velocity filter, where the limits in velocity spectrum can vary with time. Further, these p-limits, based on velocity limits, are incorporated as part of the kernel of the forward τ–p transform and, thus, represent an alteration of the basic transform. As mentioned above, application of these p limits is the same as expanding the number of variables in which the filter function is defined. Further, it represents an inclusion of additional information, not usually included in the definition of the

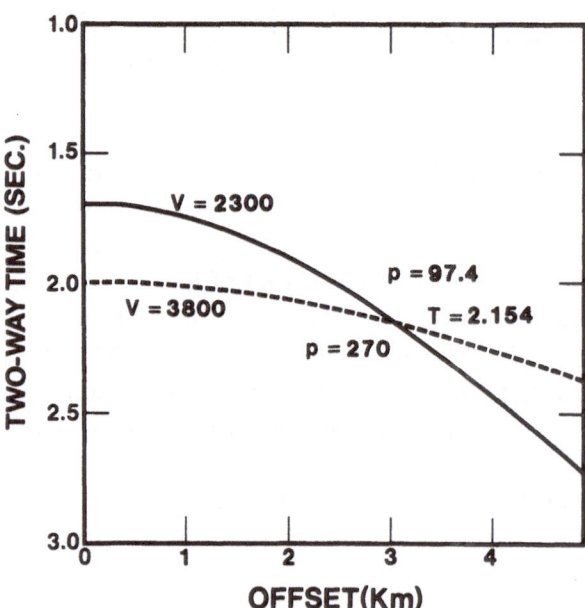

Fig. 8.   Application of reflection velocity limits in forward τ–p transform. For this example, the upper velocity limit is 3750 m/sec and lower velocity limit is 2250 m/sec (3000 ±25%). At X = 3000 m and t = 2.154 sec, we only include p-values between 97 and 270 μsec/m. At different X–t locations, the limits in p-values, even for the same velocity, will be different (after Tatham et al., 1981).

transform, into the filtering process. The details of this variation are discussed in Tatham (1984).

Examples of Hyperbolic Velocity Filtering

The effects of the hyperbolic velocity filtering resulting from the modified $\tau-p$ transform discussed above are illustrated in Figures 9 and 10. The data are the same physical model data used in the earlier examples. The limits in velocity (shown in Table I) are consistent with *P*-wave velocities, ±25%, for Figure 9 and are consistent with shear-wave velocities, ±25%, for Figure 10. Comparison with the full $\tau-p$ transform shown in Figure 4 shows the full effects of the hyperbolic filtering. Again, the filtering is accomplished by altering the kernel of the forward transform, not by simply band-limiting the data in transform space. This is illustrated by the improved continuity of the reflection ellipses in Figures 9 and 10 over those in Figure 4. The effect is particularly pronounced in the shear-wave

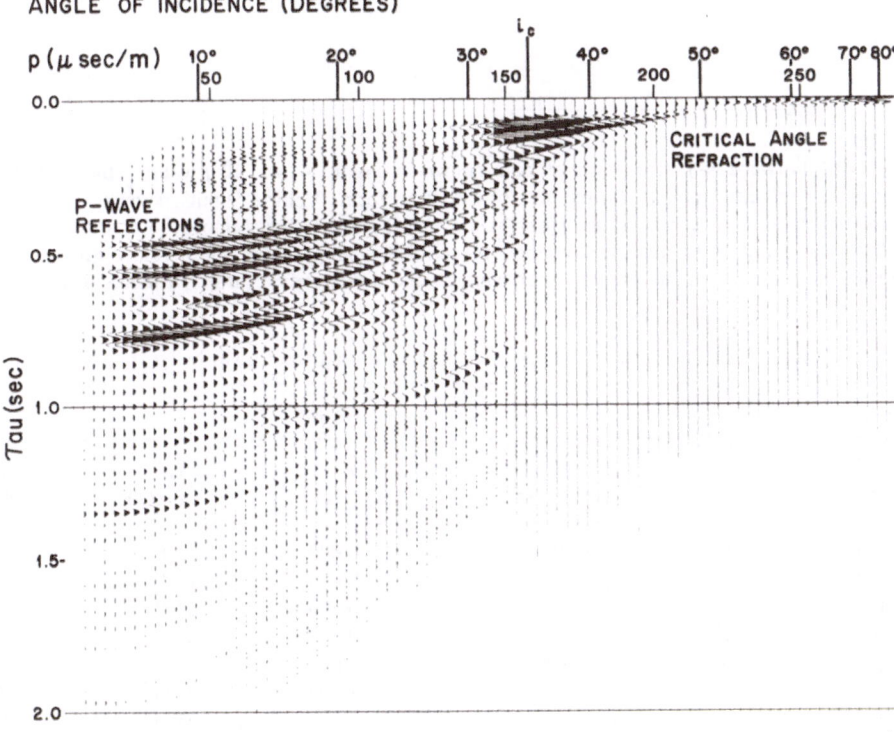

Fig. 9. Forward $\tau-p$ transform of the data shown in Figure 3, for all *p*-values, but with limits in hyperbolic velocity applied. Time-varying hyperbolic velocity limits, given in Table I, are consistent with *P*-wave reflections. Limits in hyperbolic velocity are applied through a modified kernel during the forward $\tau-p$ transform. Comparison with Figure 4 shows improved continuity of reflection ellipses for *P*-wave reflections as well as reduction in noise in areas of transform not consistent with *P*-wave reflections (after Tatham, 1984).

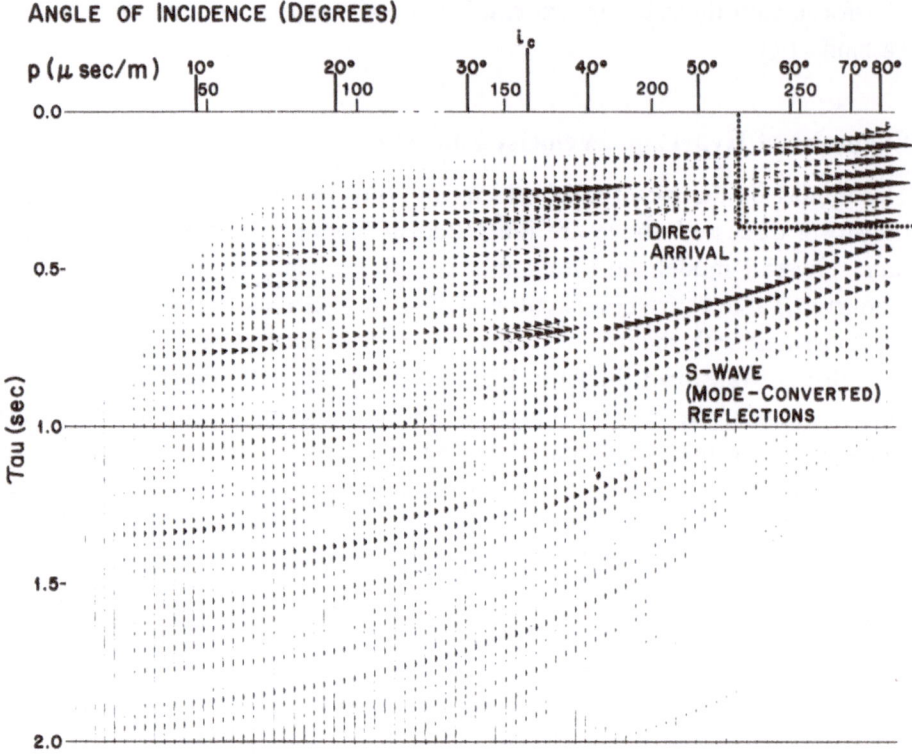

Fig. 10. Forward $\tau-p$ transform of data shown in Figure 4, for all p-values, but with limits in hyperbolic velocity applied during the forward transform. Velocity limits, shown in Table I, are consistent with S-wave reflections. Comparison with Figure 4 shows improved continuity of S-wave reflections at angles of incidence beyond the critical angle. Region for p-values greater than 230 μsec/m and time less than about 0.3 sec contains the direct arrival (after Tatham, 1984).

Table I.   Limits in NMO reflection velocities applied during the forward $\tau-p$ transform

P-wave			S-wave		
T_0 (sec)	V_{low} (m/sec)	V_{high} (m/sec)	T_0 (sec)	V_{low} (m/sec)	V_{high} (m/sec)
0.000	2750	4575	0.000	1350	2250
0.020	2750	4575	0.040	1350	2250
0.100	4475	4725	0.200	2200	3650
0.200	4650	7775	0.400	2300	3850
3.000	4800	8000	1.000	2350	3900
			1.800	2400	4000
			3.000	2500	4200

transform of Figure 10, especially at p-values where the angles of incidence are less than the critical angle. Note that S-wave reflections are present, but that the P-wave reflections are noticeably attenuated. Further, the reflections along elliptical trajectories show improved continuity with the hyperbolic filter applied.

Geophysical Examples of Applications of $\tau-p$ Filtering

GROUND ROLL

Ground roll is surface wave energy that propagates along the surface and near the surface with relatively low velocity, often with low frequency, and usually with high amplitudes relative to other seismic events of interest. The effect in the two-dimensional offset-time seismic record is a coherent high amplitude event that interferes with desired reflections. The low velocity and nearly linear character of the ground roll, however, suggest $\tau-p$ filtering as a practical and effective means of filtering such undesirable events from the $X-t$ record.

HIGH VELOCITY GROUND ROLL

As an example of the removal of ground roll, consider the seismic record shown in Figure 11 (from Tatham *et al.*, 1983b). This is a Vibroseis record from Alaska, and the relatively high velocity of the ground roll (about 1350 m/sec) often makes it difficult to filter out the ground roll without adversely affecting the desired reflections. A $\tau-p$ transform of the data (Figure 12) isolates the shot-generated ground roll in the region of p-values between 700 and 750 μsec/m and to intercept times near zero (less than 0.4 sec). The anomalous amplitudes in this region affect the trace-wise amplitude adjustment applied to this display, and is further reflected by the low amplitude at times greater than 0.4 sec. This small region, well separated from desired events, can be readily eliminated from the data input to the inverse $\tau-p$ transform, or the appropriate p-values could be eliminated from the inverse transform. Figure 13 shows the result of an inverse transform applied to the data in Figure 12, but using a range of input p-values between 0 and 625 μsec/m. That is, all coherent energy with apparent horizontal velocity less than 1600 m/sec has be omitted. Note that the undesirable ground roll is significantly diminished, with little effect on the rest of the record. By using more traces in the $\tau-p$ space than in the original record, we honor the dynamic range of the data and are able to accurately reconstruct the original record above and below the ground roll. Note the similarity of the original and filtered data in these unfiltered areas.

By placing constraints on the allowable reflection velocities, we provide additional information, independent of the seismic data, to the system. A range of normal moveout velocities for the region, known from other sources of regional knowledge, is shown in Figure 14. Note that this is a fairly broad range of permissible velocities, is applicable to many sedimentary basins, and is thus not very restrictive. Applying a modified $\tau-p$ transform incorporating these velocity limits yields the plot shown in Figure 15. Comparing the reflection ellipses at lower p-values (0 to 200 μsec/m) with those in the conventional transform shown in Figure 12 shows the improved character of the filtered reflection ellipses. Note that the ground roll information falls entirely outside the range of permissible

FIELD RECORD

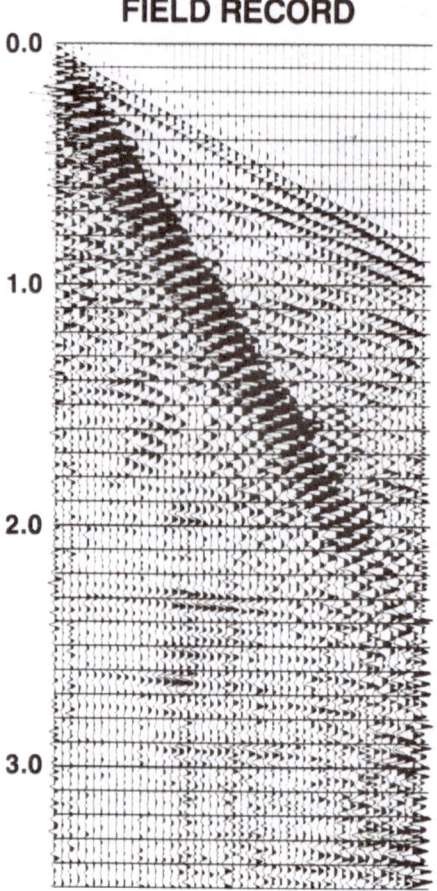

Fig. 11.   Record with high velocity (1370 m/sec) ground roll and clear reflecting at depth. This Vibroseis record is from Alaska (after Tatham *et al.*, 1983b).

velocities and is not present in this transform. In fact, no transformed data are observed at *p*-values beyond about 200 μsec/m. Application of the inverse $\tau-p$ transform to this data set yields the filtered record shown in Figure 16. Not only is the ground roll greatly attenuated, but the hyperbolic reflection character along reflections hyperbola is improved. This is a result of application of a time-varying hyperbolic velocity filtering, a technique that has proved to be of great utility in many applications.

GROUND ROLL AND A DIRECT AIR WAVE

Another example of filtering of linear events — both ground roll and a direct, low velocity, air wave — is provided by a data set from the Ouachita uplift area. In this example, shown in Figure 17, two separate linear trending events interfere with the desired hyperbolic reflections. The forward $\tau-p$ transform, illustrated in

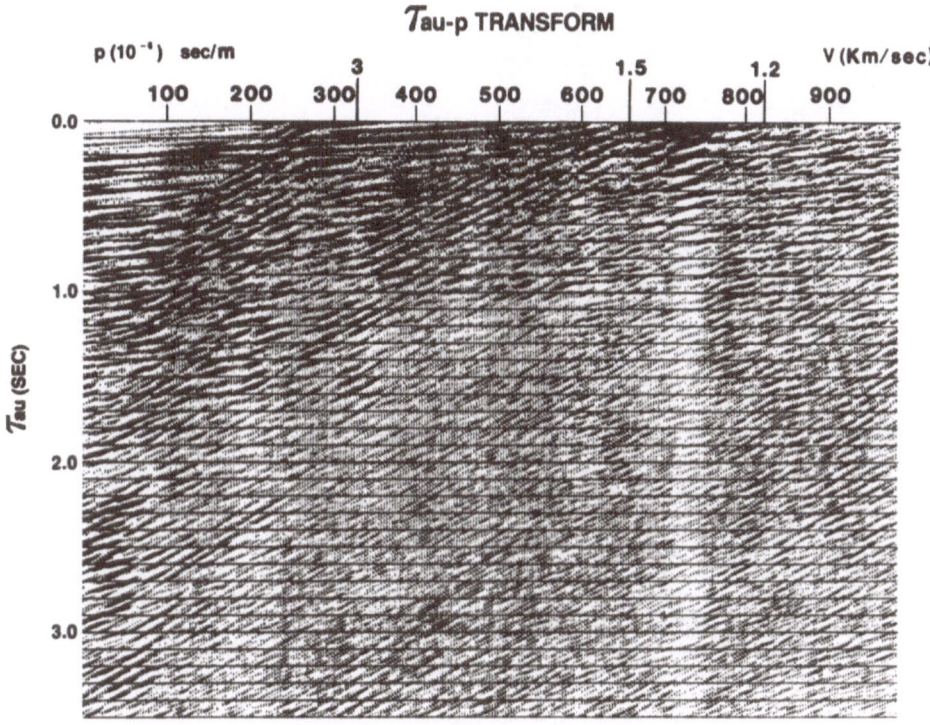

Fig. 12.   Forward $\tau-p$ transform of record in Figure 11. Note the concentration of amplitude near $\tau=0$ and $p=720$ (1370 m/sec). Low amplitude below the zone of interest results from large amplitude at $t<0.4$ sec and trace-wise display normalization (after Tatham *et al.*, 1983b).

Figure 18, concentrates the ground roll near a point at about 460 μsec/m and $\tau = 0$ sec, but the very slow air wave is beyond the range of *p*-values included in the transform. The reflection events are generally restricted to a range of *p*-values between 0 and about 130 μsec/m.

Applying the hyperbolic reflection velocity filter during the transform process, with constant limits of 3000 to 9000 m/sec, yields the modified transform shown in Figure 19. Note the improved continuity in the reflection ellipses in the range of *p*-values 0–130 μsec/m. A conventional inverse $\tau-p$ transform applied to the data of Figure 19 yields the filtered field record shown in Figure 20. Comparison of this record with the original record shown in Figure 17 demonstrates the improvement realized just through the application of the reflection velocity limits as part of the forward $\tau-p$ transform. As will be discussed below, other filtering could also be applied in the τ domain prior to inverse transform and further enhance the results.

FILTERING FOR COHERENT SCATTERED ENERGY

One source of coherent noise on seismic field records is the scattering of source-generated signals. For example, in a marine area with a hard water bottom (i.e.,

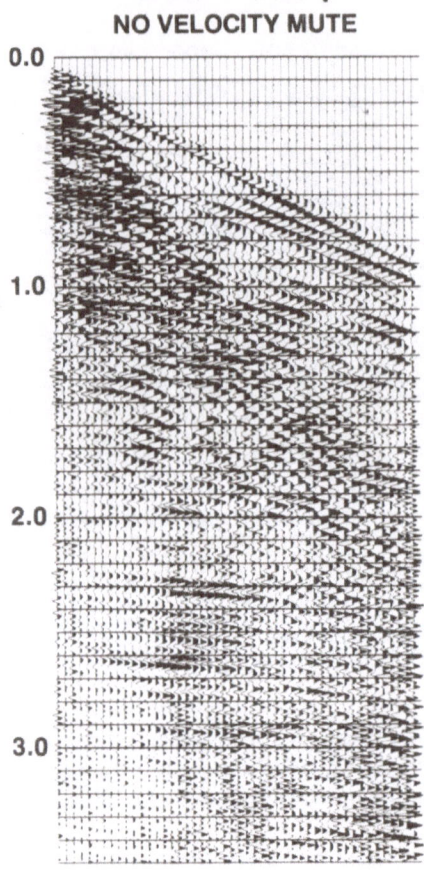

AFTER $\mathcal{T}$au-p

NO VELOCITY MUTE

Fig. 13. Inverse transform of $\tau-p$ data shown in Figure 12. Range of p-values was 0 to 620 μsec/m, eliminating the air-coupled ground roll. The similarity of the transformed data and the original data above and below the zone of the ground roll demonstrates the robustness of the procedure (after Tatham *et al.*, 1983b).

an effective channel for the propagation of scattered energy) and rough topography of the water bottom (i.e., many sources of scattering), water-borne energy scattered from ahead, to the side of, and behind the recording are commonly observed. (See, for example, Larner *et al.* (1983) for a rather complete discussion of this problem.) Such energy is quite coherent and propagates with the velocity near that of acoustic waves in water. The apparent velocity along the recording array, however, may be significantly different, depending on whether the scattered energy is ahead of, to the side of, or behind the recording array. Further, the zero intercept time is not, in general, the same as the source initiation time (as is the case with ground roll).

Larner *et al.* (1983) suggest conventional $f-k$ filtering as a means of attenuating

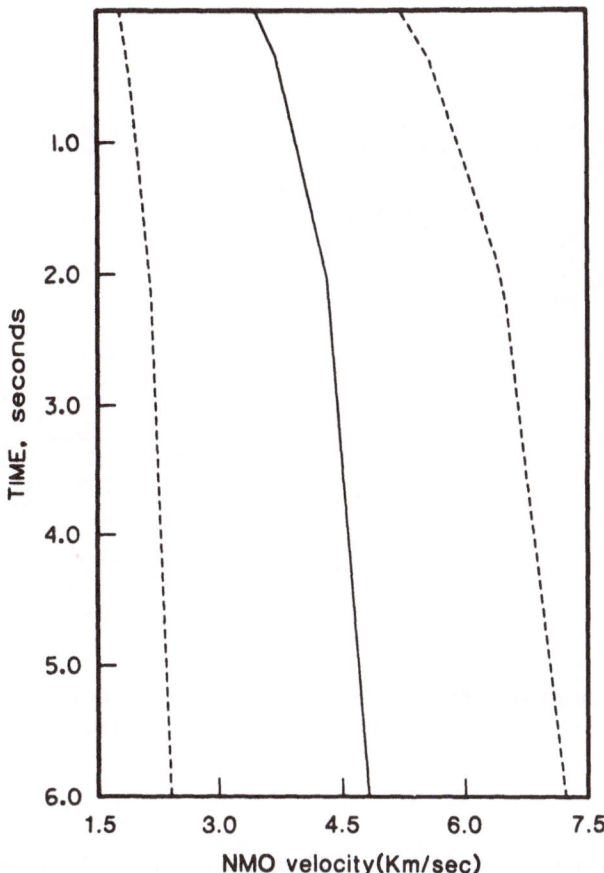

Fig. 14.   Velocity-time functions describing the limits of NMO reflection velocities to apply during forward τ–p transform. The solid line represents the stacking velocity function, and the dashed lines ($\pm50\%$) are the reflection velocity limits applied during forward transform (after Tatham *et al.*, 1983b).

such undesirable coherent energy. They suggest filtering the shot record, sorting the data to a common receiver format, and *f*–*k* filtering again in that *X*–*t* domain. Hyperbolic velocity filtering which we have been discussing, has the advantage of addressing the desired events as reflection hyperbolas and passing desirable data by that criteria. Thus, selective application of hyperbolic velocity filtering may have advantages over conventional *f*–*k* linear velocity filtering.

Noponen and Keeney (1986) show an example of such filtering. In theoretical calculations, they show that applying the hyperbolic velocity filter to the shot-ordered field records alone is equivalent to (1) *f*–*k* filtering the shot ordered record, (2) resorting the data to common receiver gathers, (3) applying *f*–*k* filtering a second time, and (4) again resorting the data to a common midpoint gather. The computer time required to accomplish all this resorting of the data, as well as the numerous applications of multi-dimensional filtering, clearly estab-

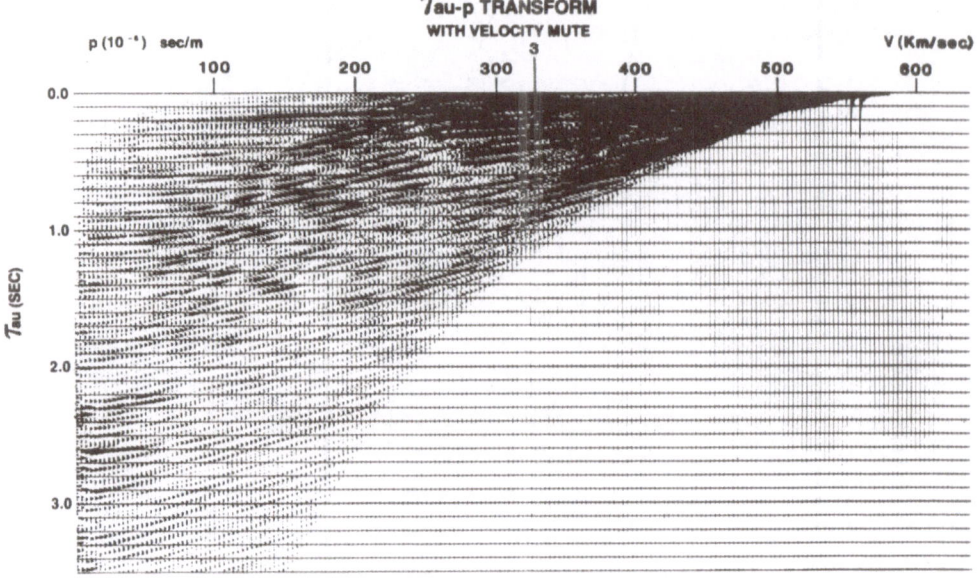

Fig. 15. Tau-*p* transform of data shown in Figure 11, but with reflection velocity limits (shown in Figure 14) applied. Note the improved continuity of reflection ellipses at low *p* values (after Tatham *et al.*, 1983b).

lishes an economic advantage to the single application of a more sophisticated filtering procedure.

Examples of the filter response calculated by Noponen and Keeney (1986) are illustrated by array response plots. Figure 21 shows the effect of CMP stacking alone, where the distance ahead of or behind the recording array is considered, as well as the distance off to the side of the recording array. Notice that the CMP stack has little effect in filtering coherent events scattered from points off to the side of the recording array. Figure 22 shows the effect of hyperbolic velocity-filtering the shot-ordered records, combined with the effect of the CMP stack. Figure 23 shows a similar effect of applying conventional linear $f-k$ filtering on both shot records and common receiver gathers. From these plots, it appears that the single hyperbolic velocity filter applied to the shot-ordered records is more effective at attenuating scattered energy than the twice $f-k$ filtered (and twice sorted) data. The results of the final stack on these somewhat poor data, however, show a more nearly-equivalent effect of the two methods than do the calculated responses. Thus, $\tau-p$ filtering with velocity limits applied is at least as effective, when applied to poor data, as $f-k$ filtering applied in both the shot and receiver gathers for attenuation of scattered energy. On higher quality data, the difference should be even more pronounced.

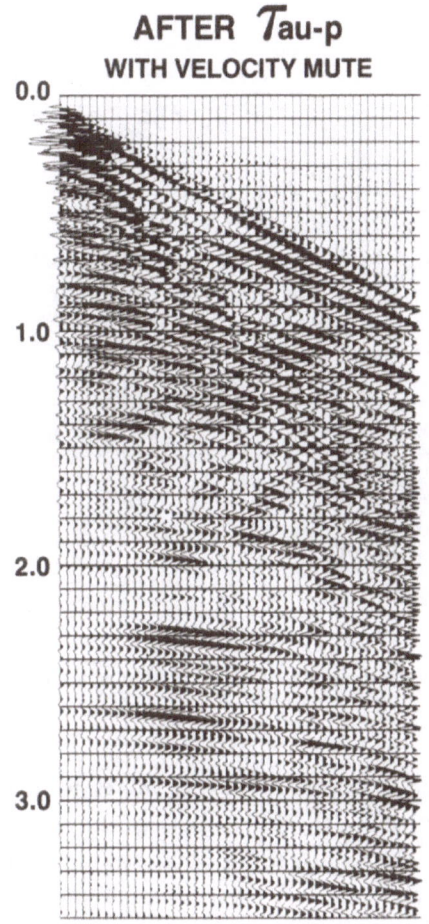

AFTER $\mathcal{T}$au-p
WITH VELOCITY MUTE

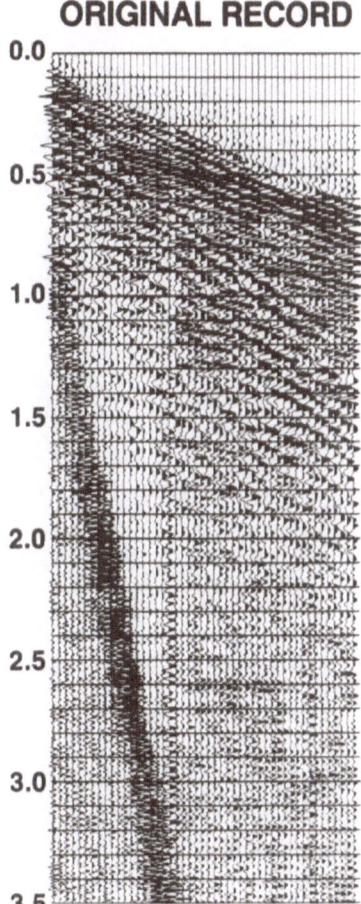

ORIGINAL RECORD

Fig. 16. Inverse transform applied to the velocity-limited transform shown in Figure 15. Comparison with Figures 11 and 13 shows an improved reflection character (after Tatham *et al.*, 1983b).

Fig. 17. Field record from Ouachita uplift area showing both ground roll and a direct air wave (after Tatham *et al.*, 1983b).

SEPARATION OF *P* AND *SV* ENERGY BY $\tau-p$ FILTERING

One very important application of $\tau-p$ filtering is the separation of *P*-wave and mode-converted *SV*-waves in the marine environment. Examples of data gathered in a physical model experiment were shown earlier to illustrate the effects of the $\tau-p$ transform. In particular, the separation of *P* and mode-converted *SV* reflections was realized by (1) exploiting the dependence of mode conversion upon the angle on incidence at the water bottom and (2) applying hyperbolic velocity filtering in the transform process. Application to real data from offshore Florida demonstrates that the technique does indeed work in a real exploration

p(μsec/m)

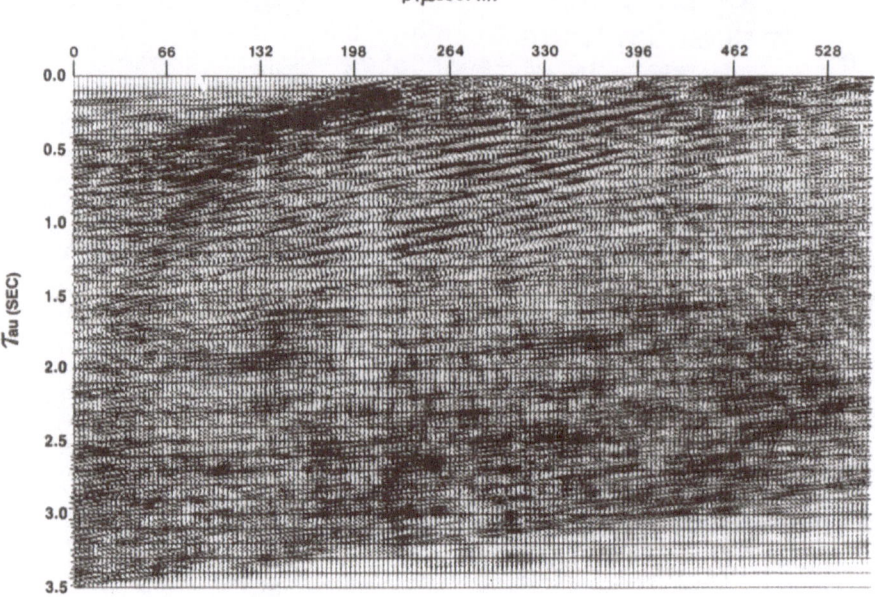

Fig. 18.   Forward $\tau-p$ transform of the field record shown in Figure 17 (after Tatham *et al.*, 1983b).

environment. Examination of the single field records, however, offers little in-sight, as the full CMP stack is required to observe the full effects of the weaker shear-wave signals. Further, offshore Florida, especially in the southern areas, has been an area of poor data quality. In fact, some of these S-wave and $\tau-p$ processed P-wave data represent the best data collected in the area.

Figure 24, from Tatham and Goolsbee (1984), shows both conventional P-wave data and a stack of the mode-converted data for a 20 mile line. Note the continuous S-wave reflections shown down to a reflection time of about 2 sec. Further, the events are correlatable between the two sections. The S-wave reflec-tion at 2 sec has a stacking velocity of about 1310 m/sec, considerably less than the propagation velocity in water. Hence, it is very difficult to imagine this event as a multiple reflection, and the apparent reflection velocity supports the event identification as a mode-converted shear wave.

Keep in mind that the mode-converted shear wave (SV) data represent two mode conversions, and that (1) the wave is generated as a P-wave in the water column, (2) mode converts to a shear wave (SV) at — or near — the water bottom, (3) is reflected from depth as a shear wave, (4) mode-converted back to a P-wave upon re-entering the water column, as (5) is recorded as a P-wave on a conventional hydrophone array. Since the wave has traveled most of its ray path as a shear wave, it has a characteristically shear wave velocity and the mode conversion introduces a strong dependence upon the angle of incidence. Thus, a logical method of separating this energy from the conventional P-wave reflections

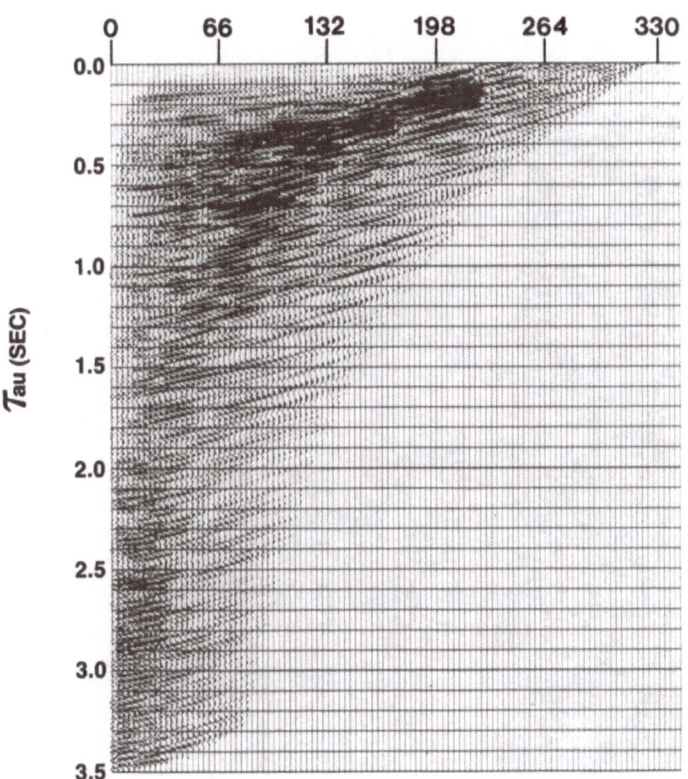

Fig. 19. Forward τ-p transform of the field record shown in Figure 17, but with velocity limits applied during the forward transform. The lower velocity limit was 3000 m/sec and the upper velocity limit was 9000 m/sec (after Tatham *et al.*, 1983b).

is the application of reflection velocity and angle of incidence criteria available through τ-p processing.

ATTENUATION OF MULTIPLES IN THE τ-p DOMAIN

In a conventional single-offset trace out of an X-t record, the time difference between successive multiples from the same multiple-generating interface decreases, even for a simple 1D earth model. Since such a time difference is not constant, multiple suppression techniques that are predicated upon periodicity of multiple energy, will not always succeed. On a constant p trace, however, multiples will be exactly periodic for a 1D earth model and, hence, most deconvolution algorithms should yield superior results over those applied on x-t records.

For example, consider the data (shown in Figure 25) from a marine area where strong multiples are a very serious problem. The similarity of the multiple and

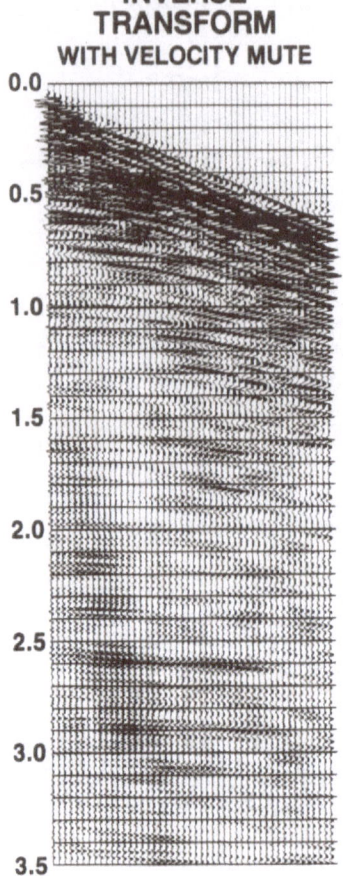

INVERSE TRANSFORM
WITH VELOCITY MUTE

Fig. 20. Inverse transform of the reflection-velocity limited transform shown in Figure 19. Comparison with Figure 17 demonstrates a pronounced improvement in the reflection character (after Tatham *et al.*, 1983b).

primary velocities minimizes the effectiveness of CMP stacking, as well as other velocity-based multiple discrimination methods, in suppressing multiple reflections. The data represent a conventional CMP gather (on the right) and the $\tau-p$ transform of that record (on the left). The $\tau-p$ transform was constructed by a method developed by Brysk *et al.* (1986) and McCowan and Brysk (this volume) which honors the cylindrical symmetry of the shot. Figure 26 represents the same data, but with a long-period predictive deconvolution applied. The deconvolution was applied to each trace, so the effects of differences in the $X-t$ and $\tau-p$ can be observed. Of course, the final results are best observed in the final CDP stack. Figure 27 shows the result of a conventional CDP stack in this area of strong multiples. The stack is of data that had a conventional deconvolution applied. The $\tau-p$ transform was applied to CDP data and, hence, after deconvolution, the data were stacked directly from the $\tau-p$ domain (i.e., no inverse transform was

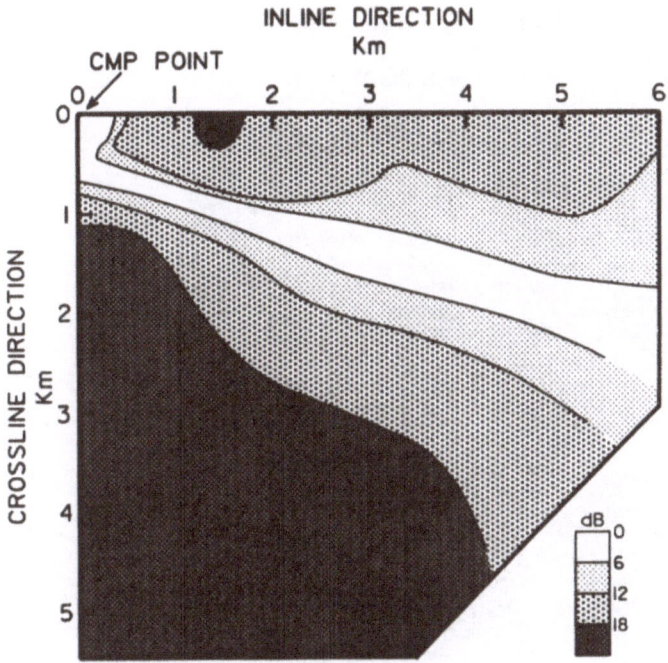

Fig. 21. Theoretical attenuation (in dB) anticipated in CDP stacking of waves diffracted by shallow scatterers relative to the CDP location and line direction. The three other quadrants have symmetrical responses. The streamer is 3562 m long and has a near-offset of 110 m (after Noponen and Keeney, 1986).

applied). The results of the $\tau-p$ deconvolution are shown in Figure 28. Comparison on Figures 27 and 28 demonstrates the effectiveness of applying deconvolution in the $\tau-p$ domain. In fact, the interpretation of the two stacked sections would likely be different. The unconformity clearly present between 1 and 2 sec in Figure 28 cannot be readily interpreted in Figure 27, and the faulting can be picked with a higher degree of confidence in Figure 28 than in Figure 27. The overall improvement in data quality is quite remarkable.

Consideration of the Effects of Dip on the Transform Process

One question that often arises when considering $\tau-p$ filtering, especially when posed in terms of a slant stack, is "What are the effects of dip?" The same question also arises with respect to the effects of the hyperbolic velocity filter. The question revolves around the position of the apex of the reflection hyperbola, not necessarily at zero offset on the shot-ordered field records. Dip on a plane reflector causes a translation of the reflection hyperbola to a position where the apex of the hyperbola is at some finite source-receiver offset on the field record. This effect is especially apparent in split-spread shooting configurations. To some degree, the wide ranges of velocity used in the forward $\tau-p$ transform will

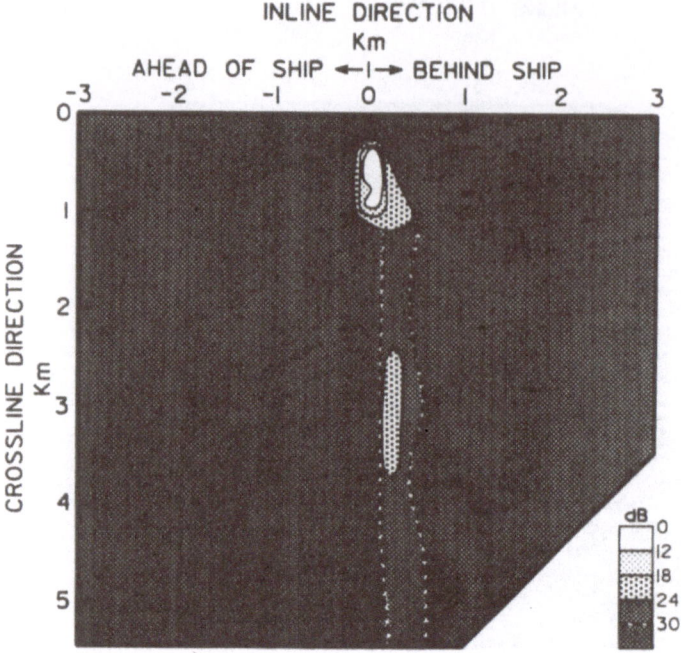

Fig. 22. Theoretical attenuation anticipated in CDP stacking *combined* with the hyperbolic velocity filtering of shot records when stacking waves are diffracted by shallow scatterers. The attenuation is shown as a function of the position of the scatterer relative to the position of the ship (source). We assume a 3562 m long streamer and a ship traversing from right to left. The other half-plane has an identical response (after Noponen and Keeney, 1986).

compensate for a range of slopes. It is possible, however, for the reverse slopes to appear on the record. Thus, we need to allow for some negative p values where a large dip is anticipated.

Figure 29 shows a split-spread field record collected in the Anadarko Basin of Oklahoma. Note the asymmetry of some of the reflections' hyperbolas, especially the event near 2.7 sec. Forward $\tau-p$ transforms, with reflection velocity limits between 2500 and 8000 m/sec, were applied to this record. Figure 30 shows the results where the negative p values are derived only from the left side (negative X) of the record and the positive p values from the right side of the record. The inverse transform, shown in Figure 28b, does not honor the overall characters of the reflection hyperbolas observed in the original record. When negative p values are included for the right side of the record and positive p values for the left side, we honor the reflection hyperbolas, as shown in Figure 31. The actual dip angle corresponding to these p values is dependent on the true velocity of the stratigraphic section, which is not always known precisely. For this example, however, the 100 μsec/m range of inverse slope corresponds to a range of dip angles everywhere greater than ±15 degrees. The $\tau-p$ transform using this allowance is shown in Figure 29c. Note that the reflection ellipses are translated in the $\tau-p$

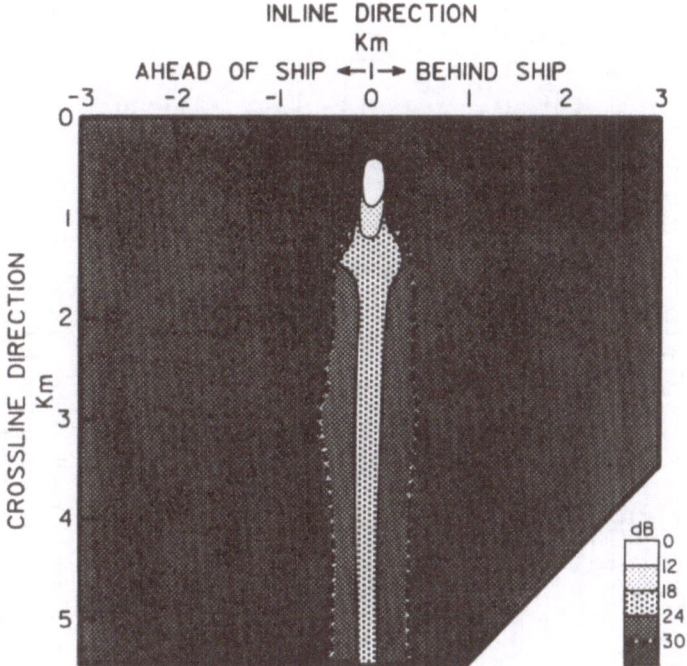

Fig. 23. Theoretical attenuation anticipated in CDP stacking *combined* with conventional *f–k* velocity filtering applied to *both* shot and receiver gather, when stacking waves are diffracted by shallow scatterers. The attenuation is shown as a function of the position of the scatterer relative to the position of the ship (source). Comparison with the results in Figure 22 shows less attenuation with two passes of the *f–k* filter and a single pass of the velocity limited τ–p filter (after Noponen and Keeney, 1986).

space just as reflection hyperbolas are translated in the X–t space. The application of the reflection velocity limits have improved the reflection character, and the allowance for dip has honored the original hyperbolas slopes.

The reliability of the transform, even when dipping events are present, will be useful in future applications, especially where imaging applications are anticipated. The plane wave nature of the data as it emerges at the surface of the earth, is fully preserved by the τ–p transform, and many of the filtering processes remain valid, even in areas of complex structure.

Review of Examples

In this discussion, I have attempted to illustrate real-data examples of multidimensional filtering by application of the τ–p transform. More effective filtering can be realized by employing a hyperbolic velocity filter as part of the transform process.

Examples include attenuation of ground roll, separation of *P* and *SV* energy in a marine environment, attenuation of scattered shot-generated energy from

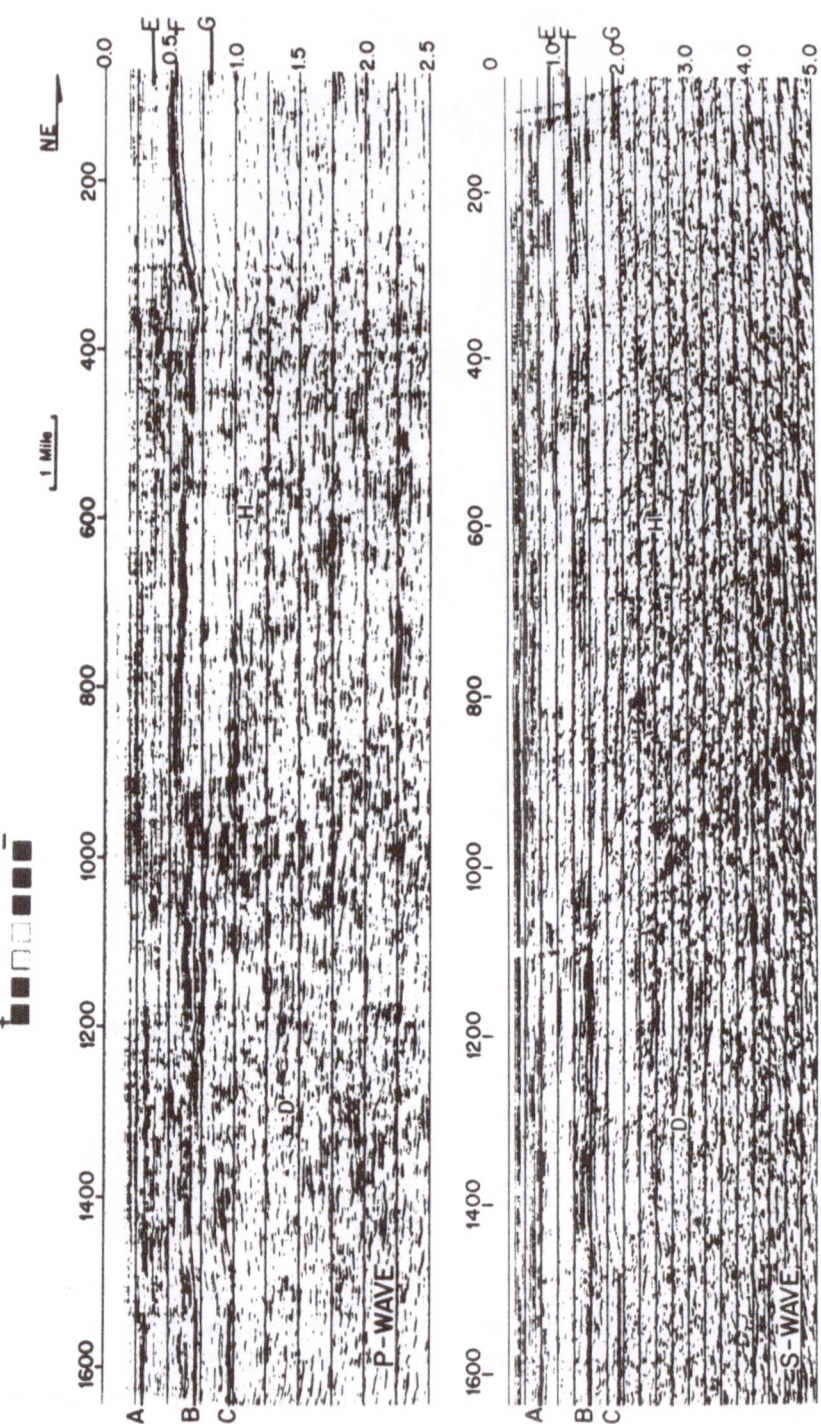

Fig. 24. *P* and *S*-wave data with τ–*p* processing applied. The data were collected along a common profile offshore southwest Florida, an area of hard-water-bottom conditions and generally poor data quality. The top section is *P*-wave data, while the lower portion is the *S*-wave section. *P* and *S*-wave reflections were separated by τ–*p* processing by pronounced differences in stacking velocity. The *S*-wave section is at a time scale one-half that of the *P*-wave section, compensating for the nearly two-to-one velocity difference in *P* and *S*-wave velocities. Note the structural correlation to 2 sec *S*-wave time. Letters indicate the correlation of reflectors between the two data sets (after Tatham and Goolsbee, 1984).

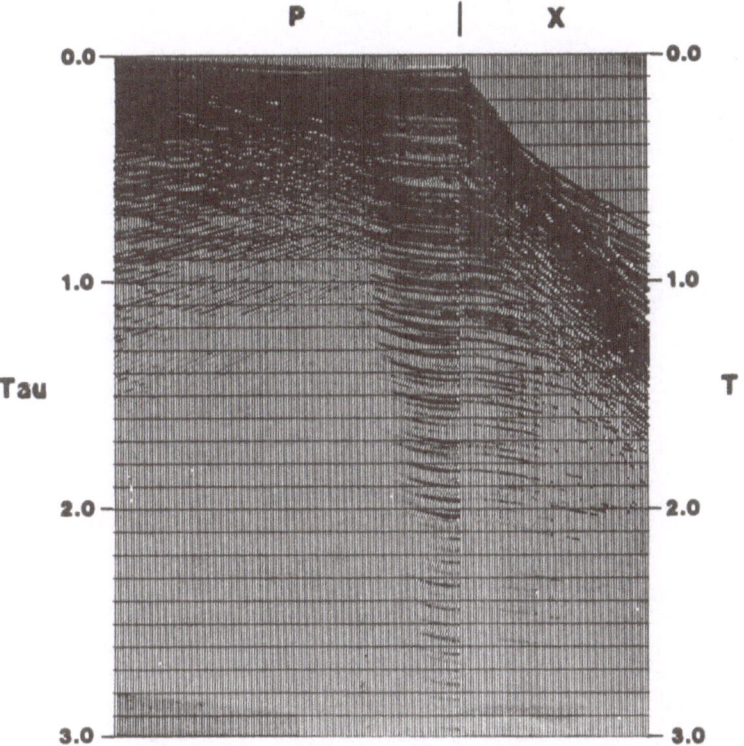

Fig. 25. A single CDP record and associated $\tau-p$ transform. This area displays strong multiple energy. This single display, courtesy of CGG, shows the relation between the field record and the $X-t$ data.

shallow-point scatterers and multiple attenuation through appropriate deconvolution in the $\tau-p$ domain.

Future applications of $\tau-p$ filtering will be in areas of strong, coherent noise. Such problems exist in many marine areas, especially those areas with hard water bottoms and irregular bathymetry which can act as a source of scattered energy. Tau-p filtering shows special promise in processing of data recorded with long marine cables, since it separates the data so well and exploits all of the information recorded at large offsets. In land settings, sources of scattering of near surface noise include clinker beds, and areas of near-surface volcanics. Considerable success has been demonstrated by some processors in filtering of near-surface scattered energy from clinker beds, and further experimentation in areas of near-surface volcanics should be encouraged.

Acknowledgements

I wish to thank Texaco for the support and encouragement in preparing this paper. Dr T-C. Shih and Ms R. B. Oliveros offered critical reviews of the manu-

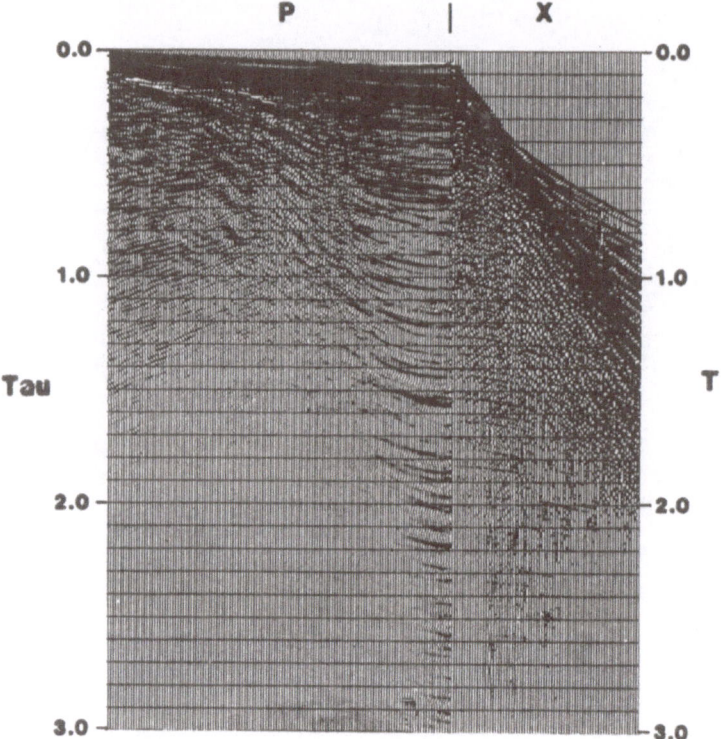

Fig. 26. Same data as shown in Figure 24, but with a long-period gapped predictive deconvolution applied to attenuated long-period multiples. The deconvolution was applied to both the $\tau-p$ and the $X-t$ records (processing courtesy of CGG).

script, and Dr P. M. Krail was instrumental in preparing the multiple attenuation portion of the discussion. Dr L. M. Jackson assisted with her editorial skills.

References

Alam, A. and Austin, J., 1981, Suppression of multiples using slant stacks, presented at the 51st annual SEG meeting, Los Angeles.

Brysk, H., Goodrum, R., and Mohan, M., 1986, Alternative implementation of cylindrical $\tau-p$, presented at the 56th annual SEG meeting, Houston.

Larner, K., Chambers, R., Yang, M., Lynn, W., and Wai, W., 1983, Coherent noise in marine seismic data, *Geophysics* **48**, 854–866.

Noponen, I. and Keeney, J., 1986, Attenuation of waterborne coherent noise by application of hyperbolic velocity filtering during the Tau-p transform, *Geophysics* **51**, 20–33.

Phinney, R. A., 1981, Wide-angle reflection seismology, presented at the 51st annual SEG meeting, Los Angeles.

Phinney, R. A., Chowdhury, K. R., and Frazer, L. N., 1981, Transformation and analysis of record sections, *J. Geophys. Res.* **86**, 359–377.

Schultz, P. S. and Claerbout, J. F., 1978, Velocity estimation and downard continuation by wavefront synthesis, *Geophysics* **43**, 691–714.

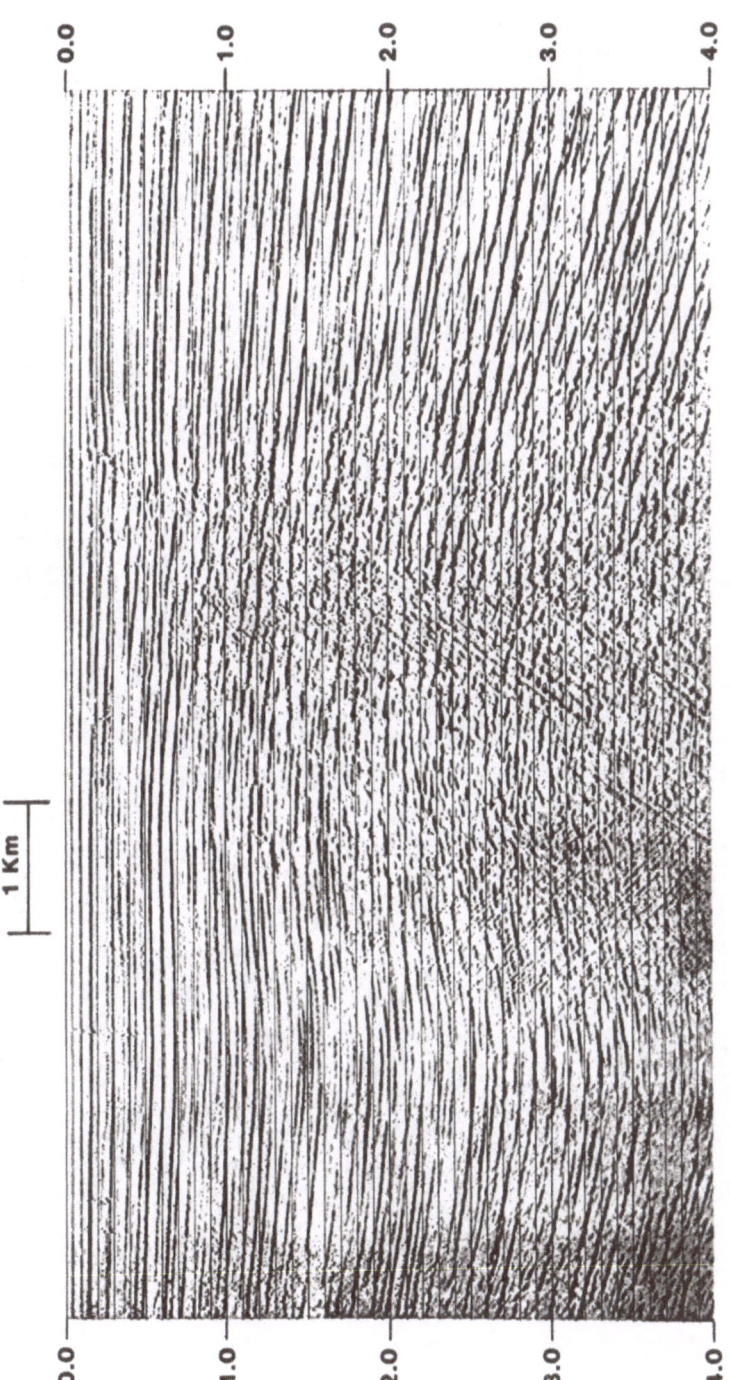

Fig. 27. The common depth point stack of an entire line of X–t records like those shown in Figure 25. This is a conventional stack and shows the effect of numerous multiples (processing courtesy of CGG).

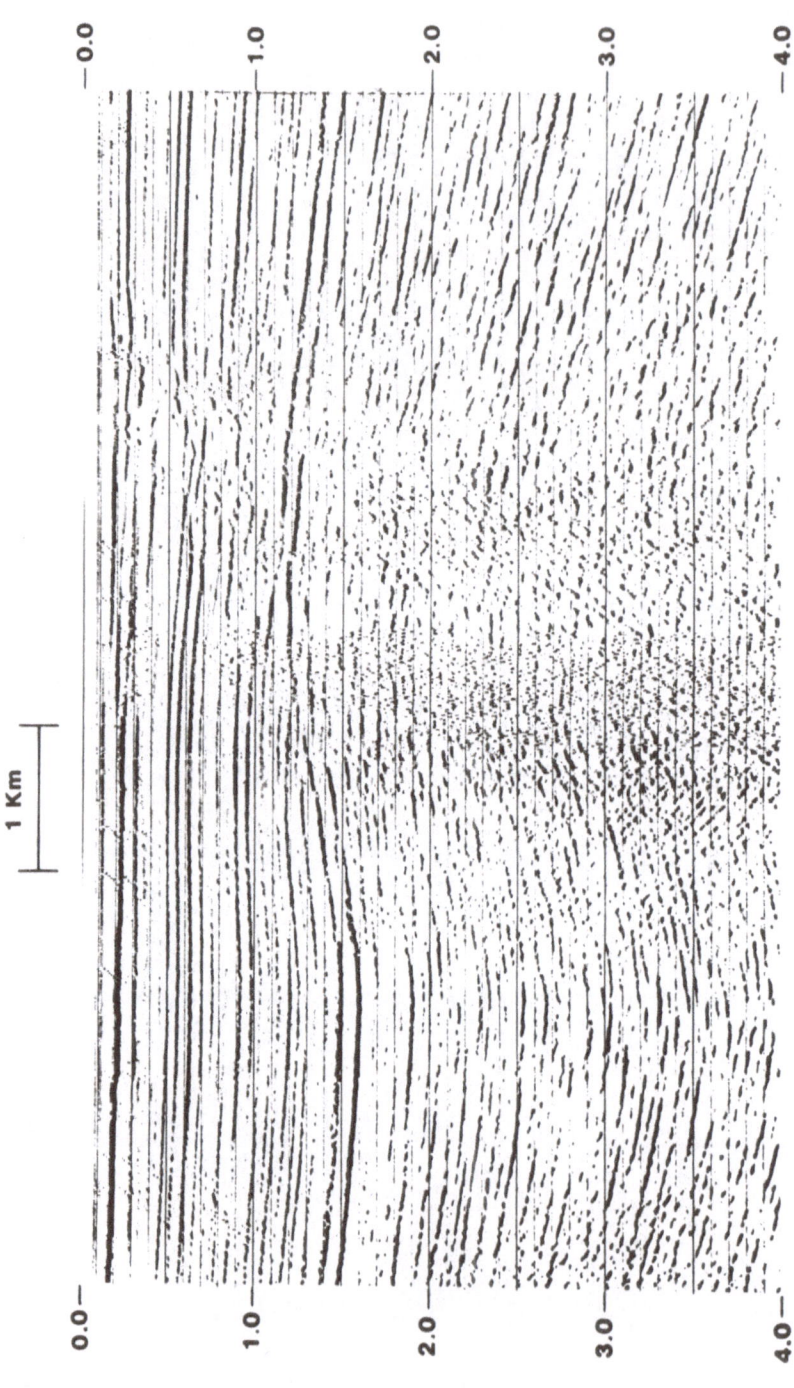

Fig. 28. The stack of $\tau-p$ records on an entire line of data like that shown in Figure 25. The difference between this record section and that in Figure 26 is that the deconvolution was applied in the $\tau-p$ domain. Stacking was along elliptical trajectories in the $\tau-p$ domain (processing courtesy of CGG).

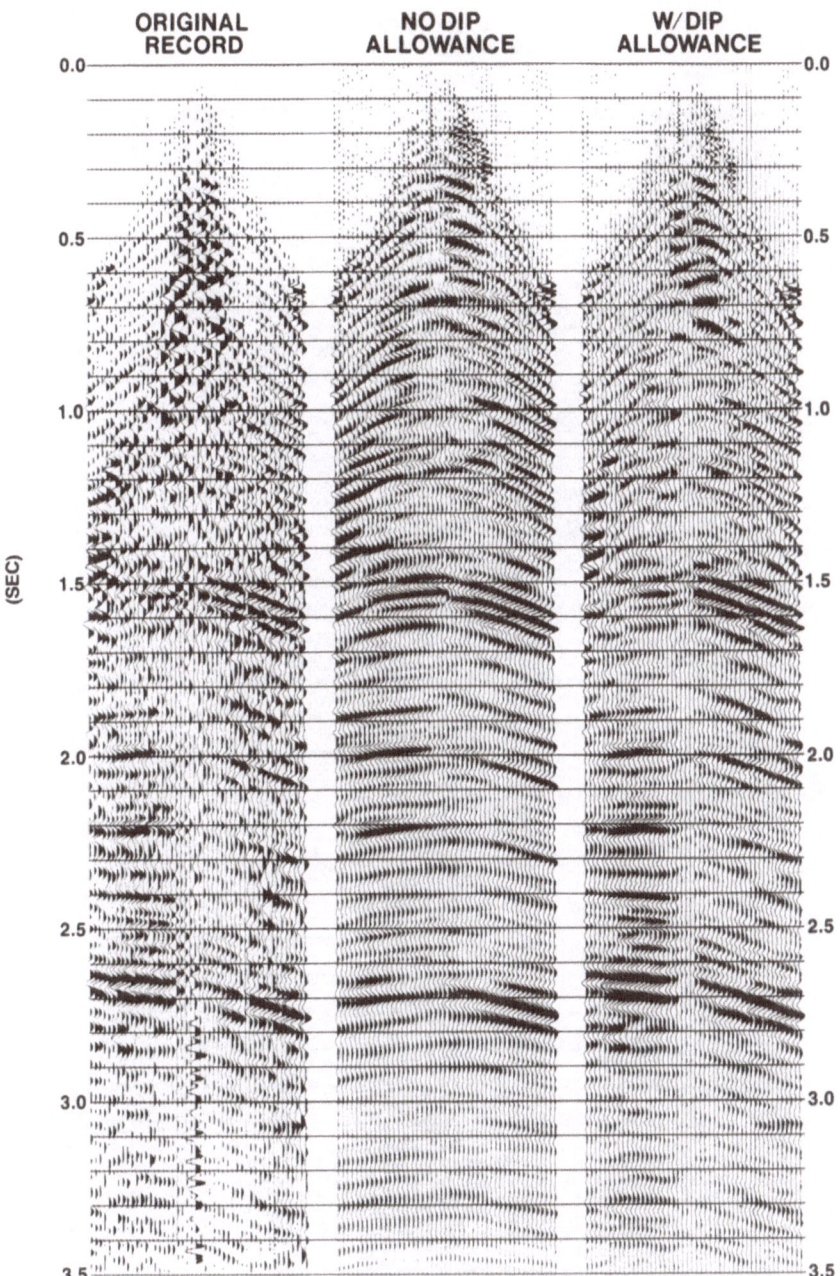

Fig. 29. The left panel is a split-spread field record from the Anadarko Basin of Oklahoma. The second panel (center) is the same data after transformation to $\tau-p$ and inverse transform back to record space, without any consideration for structural dip. The third panel (right) represents a similar pair of transforms, but with the dip allowance included in the forward transform. Note how well the right panel honors the original reflector slope, especially near $t = 2.7$ sec (after Tatham *et al.*, 1983b).

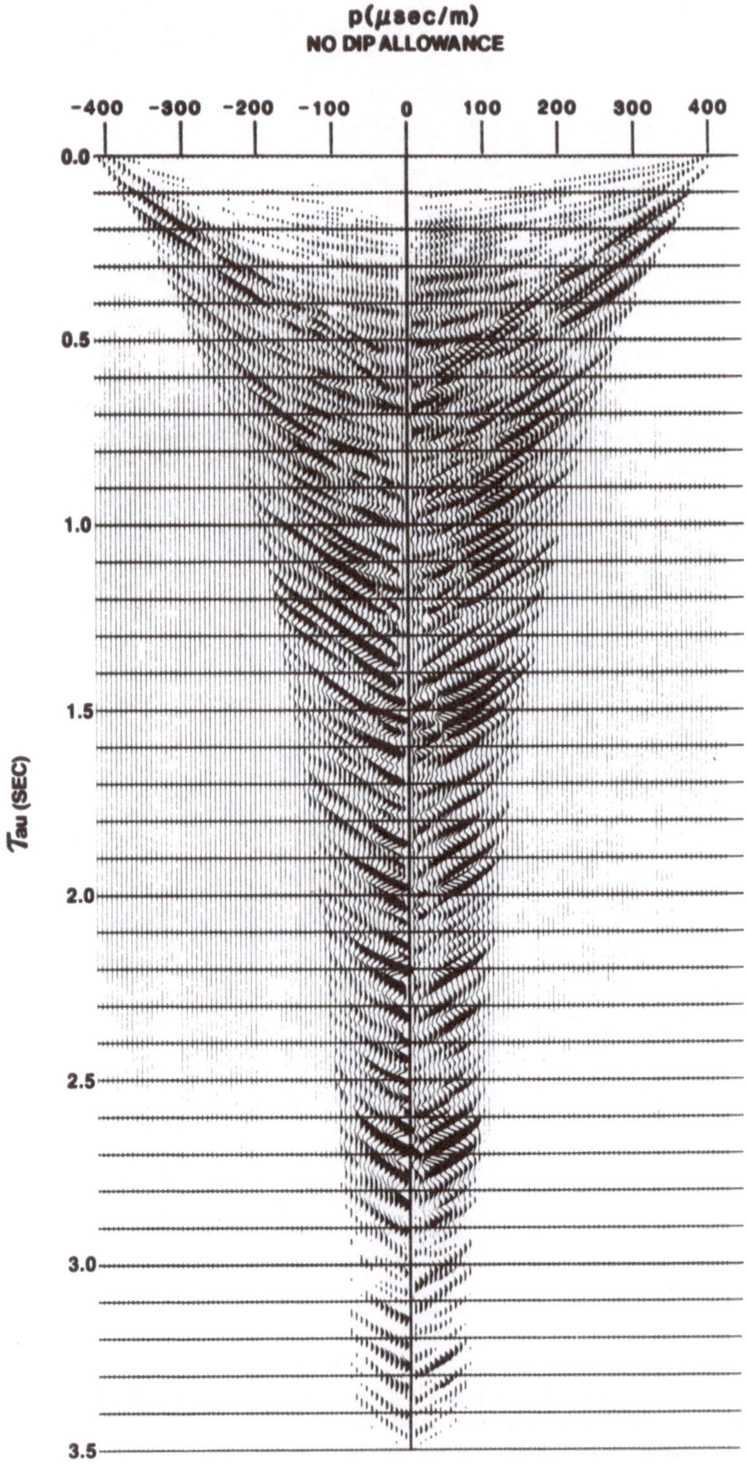

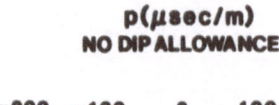

Fig. 30.   Tau-*p* transform of the split-spread record shown in Figure 29. No dip correction has been considered. Reflection velocity limits of 2500 to 8000 m/sec have been applied (after Tatham *et al.*, 1983b).

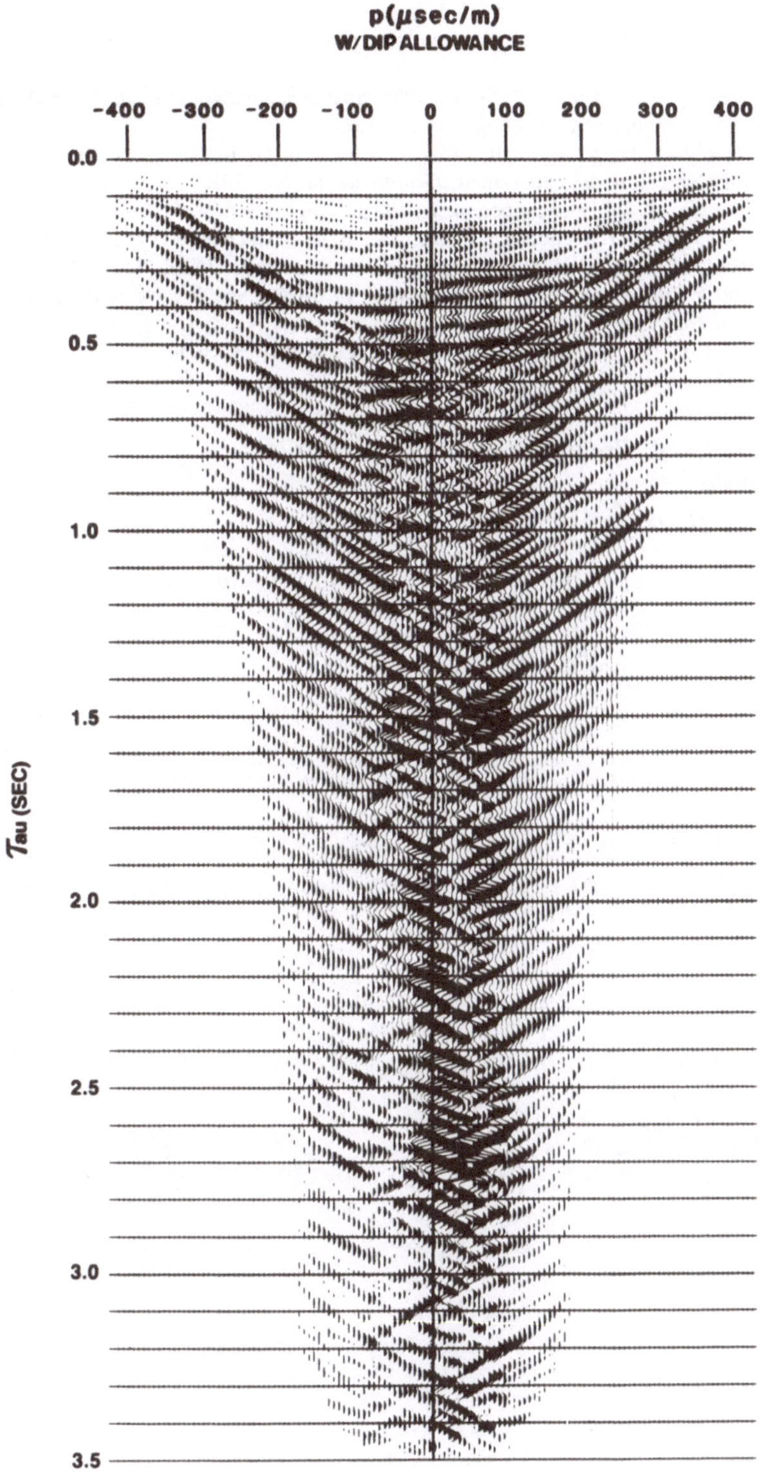

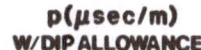

p(μsec/m)
W/DIP ALLOWANCE

Tau (SEC)

Fig. 31.   Tau-*p* transform of the record shown in Figure 29, with dip allowance applied. Note the shift in symmetry from *p* = 0 to positive *p* values. See the text for a discussion of dip allowance (after Tatham *et al.*, 1983b).

Stoffa, P. L., Buhl, P., Diebold, J. B., and Friedemann, W., 1981, Direct mapping of seismic data to the domain of intercept time and ray parameters, a plane-wave decomposition, *Geophysics* **46**, 255–267.

Tatham, R. H., Kenney, J., and Massell, W. F., 1981, Spatial sampling and realizable 3-D surveys, presented at the 51st annual SEG meeting, Los Angeles.

Tatham, R. H., Goolsbee, D. V., Massell, W. F., and Nelson, H. R., 1983a, Seismic shear-wave observations in a physical model experiment, *Geophysics* **48**, 688–701.

Tatham, R. H., Keeney, J. W., and Noponen, I., 1983b, Application of the $\tau-p$ transform (slant-stack) in processing seismic reflection data, *Aust. Soc. Explor. Geophysics* **14**, 163–172.

Tatham, R. H., 1984, Multidimensional filtering of seismic data, *Proc. IEEE*, **72**, 1357–1369.

Tatham, R. H. and Goolsbee, D. V., 1984, Separation of S-wave and P-wave reflections offshore western Florida, *Geophysics* **49**, 493–508.

Treitel, S., Gutowski, P. R., and Wagner, D. E., 1982, Plane wave decomposition of seismograms, *Geophysics* **47**, 1375–1401.

Texaco Exploration & Producing Technology Division,
P.O. Box 770070,
Houston, TX 77215-0070,
U.S.A.

JOHN DIEBOLD

Tau-p Analysis in One, Two and Three Dimensions

Introduction

Seismic inversion methods that are based on the analysis of traveltime rely on the most fundamental measurement of waveform and amplitude; either there *is* an arrival or there *is not*. Traveltime contains the basic information about an arrival's phase; the geometrical arrival time is equal to the stationary point of the phase term in the wave equation representation for any seismic arrival. Even when impedance and velocity series are determined from the amplitude and phase variations along a seismic trace, these can usually be interpreted only as fine scale deviations from a velocity function which changes relatively slowly as a function of depth. That 'long period' velocity function must itself be obtained by other means; typically from traveltime analysis.

Exploration seismic profiles designed to record traveltimes for the purpose of velocity analysis have traditionally made use of horizontally arrayed shot and/or receiver points. The resulting data comprise a set of seismograms, each providing a series of arrival amplitudes varying as a function of time (T), at a particular source–receiver offset (X). Except in a homogeneous earth, there is no simple, linear relationship directly connecting these $T(X)$ data with the desired velocity-depth ($V(Z)$) function.

Reflected arrivals have traditionally been analyzed with methods that treat each reflection as if it were returning from the bottom of a single, homogeneous layer. $V(Z)$ would be obtained later, by using Dix's (1955) method, which compares the homogeneous velocities obtained from a series of reflections. Refracted arrivals were usually treated in a completely different way. The arrival's slopes ($\mathrm{d}T/\mathrm{d}X$) and intercepts τ would be graphically determined, and under the assumption that the arrivals in question were headwaves refracted at the interfaces between homogeneous layers, the intercepts would be used to solve the layer thicknesses. The velocities, of course, were derived from the slopes. Many other methods were developed and used, and often the results from the reflected wavefield were later combined and tabulated along with those from the refractions, but there was no true unification of these two (apparently very different) groups of techniques and the results were often quite crude, owing to the time-intensive nature of the analysis.

The innovation of the concepts and mechanics of the $\tau(p)$ transform has

Paul L. Stoffa (ed.), Tau-p: A Plane Wave Approach to the Analysis of Seismic Data, 71–117.
© 1989 *by Kluwer Academic Publishers.*

revolutionized traveltime analysis. Analagously to the way that plate tectonic concepts have allowed the unification of many previously unconnected geological phenomena, $\tau(p)$ has provided a unifying framework in which seismic arrivals of all types and the physical laws controlling them can be described and character- ized with greater clarity than ever before. The $\tau(p)$ domain is a fine place to consider and perform both forward and inverse analysis of the relationships between velocity-depth models and the corresponding observational data.

The second innovation that has enabled the development and satisfactory employment of $\tau(p)$ methods is the development of fast and capacious digital computers. In its first incarnations, the $\tau(p)$ transformation was performed via manual graphical methods that were, if anything, more tedious and time con- suming than the straight-line slope-intercept methods (in fact, as we shall see, these two approaches were nearly equivalent). Only in the last decade have various techniques been developed to perform the transformation rapidly and accurately, fully and automatically exploiting the full information content of multitrace exploration seismic data with broad offset coverage.

The intent of this chapter is to provide a simple but comprehensive description of arrival times in $\tau(p)$, and to review the various basic ways in which they may be manipulated to obtain velocity functions. Following an introductory historical review, the chapter is divided into four main sections. First, a general equation is developed to describe arrival times in simplified three-dimensional geometry. The three remaining sections are devoted to the interpretation of seismic arrivals in $\tau(p)$. Analysis methods for one-, two-, and three-dimensional velocity functions are treated in turn, each with its own section. Single-dimensional inversion is the most commonly used and the corresponding section includes a discussion of the basic limitations of the methods described.

Multi-Dimensional Traveltime Equations

The $\tau(p)$ transformation, when properly done, separates the effects of wave propagation in the vertical and horizontal directions. When the seismic slownesses of the physical media involved vary only with depth, the transformation yields results that allow rapid and accurate analysis of the wave interactions at depth. In the more typical case that the media slownesses vary in all directions, inter- pretation of the transformed data is more of a problem.

In a one-dimensional world, velocity varies only with depth, and the calcula- tion of traveltimes, intercept times, and their inversion for interval thickness and velocity is a relatively straightforward process; well known in one form or another for many decades. The obvious need to accommodate laterally variable models led historically to consideration of the next level of complexity, in which planes of velocity discontinuity were allowed to dip along a single axis. This simplification served until the acknowledgement of the third dimension was required, for inclu- sion of out-of-plane arrivals and reflections from discontinuities with crossdip.

Ewing *et al.* (1939) presented a proof of the traveltime equation for dipping layers in two dimensions and an inversion scheme for reversed profiles. In this version of the equation, layer thicknesses and ray angles were measured with respect to the interface normals. In later proofs and presentations of the equation, Adachi (1954) and Ocola (1972) measured the layer thicknesses in the vertical direction, but followed the earlier practice of measuring ray angles with respect to the layer normals. Johnson (1976) showed that measuring the angles from the vertical, as well, improved the form of the equation. In all of these proofs, the basic unit calculated for traveltime was in the ray's passage from interface to interface, and the equation was posed for a specific (fixed source or receiver, source and receiver on the surface) profile geometry. Diebold and Stoffa (1981) showed that measuring the horizontal offsets to the source and receiver separately allowed the inclusion of common midpoint geometry. More importantly, Diebold and Stoffa (1981) showed that the rigorous and continual decomposition of ray propagation times into vertical and horizontal components leads to forms of the traveltime equation that are directly applicable to the analysis of arrivals in the $\tau(p)$ domain.

By continuing in this approach, but now separating each interaction of the ray with a reflecting or refracting interface, and calculating the traveltime of the ray from one horizontal reference plane, through the interaction to another horizontal plane, a modular definition of the traveltime equation is developed. This proof is complicated only slightly by the inclusion of the third dimension, and a three-dimensional proof and traveltime equation are presented, of which the two-dimensional equation is a special case.

RAYS IN THREE DIMENSIONS

In two dimensions, the direction of a ray within a homogeneous medium can be specified by a single angle, which has traditionally been measured from the vertical. The slowness of the ray is specified by this angle, i, which can vary, and u, the slowness of the medium, which cannot, except in the case of anisotropy (we will not include the consideration of anisotropic effects in most of what follows). The ray then has a horizontal component of slowness, usually called the ray parameter and symbolized by the letter p. Since the horizontal slowness is the horizontal projection of the ray slowness, u, it can be found trigonometrically; $p = u \sin i$.

In three dimensions, it is useful to specify the ray by its direction cosines

$$\mathbf{I} = \langle A, B, C \rangle.$$

The slowness of the ray can also be described as a vector

$$\mathbf{P} = u\mathbf{I} = \langle P_x, P_y, q \rangle.$$

Here, the total horizontal slowness is

$$p = (p_x^2 + p_y^2)^{1/2} = u\mathbf{I}\cdot(\hat{\mathbf{i}} + \hat{\mathbf{j}}),$$

where $\hat{\mathbf{i}}$ and $\hat{\mathbf{j}}$ are unit vectors along the horizontal X and Y axes.

Up- and downgoing rays have vertical cosines with opposite signs, but horizontal cosines with the same sign. Horizontal cosines, however, change sign with horizontal direction, while vertical cosines do not. The result is that the incremental traveltime along a ray is always positive in the direction of the ray

$$dT = p_x\, dX + p_y\, dY + q\, dZ. \tag{1}$$

THREE-DIMENSIONAL RAY PROPAGATION

The shortest distance between any point along a ray and a planar discontinuity is in the direction of the discontinuity's normal vector. Fermat's principle and Snell's law require that the reflected and transmitted rays lie within another plane, which is defined by the incident ray and this normal vector, at the point of intersection. Figure 1 shows an incident ray, $\mathbf{I}_{a1}$, intersecting an interface whose normal is $\mathbf{N}$, giving rise to a refracted ray, $\mathbf{I}_{a2}$, and a reflected ray, $\mathbf{I}_{b1}$.

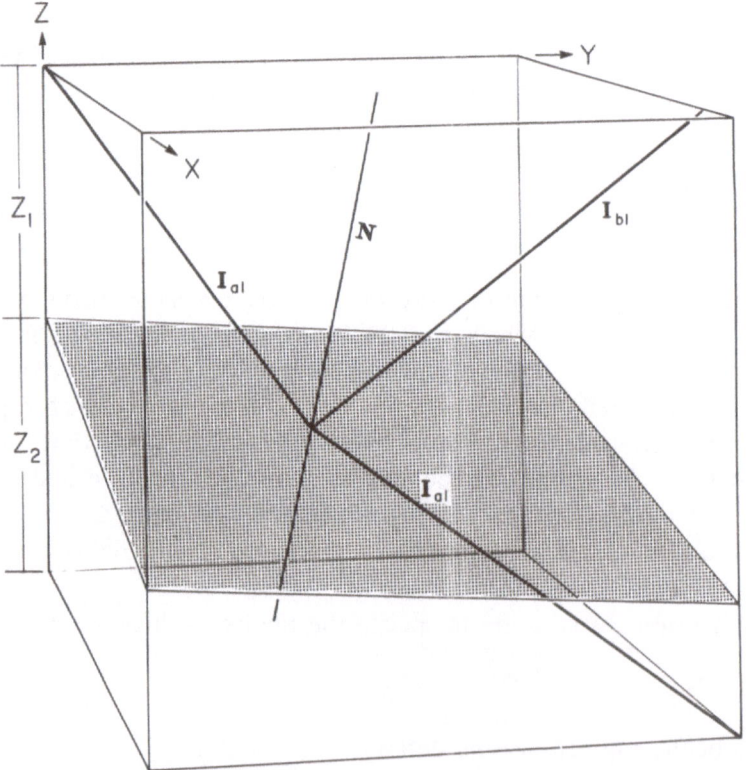

Fig. 1.   A ray, $\mathbf{I}_{a1}$ incident upon an interface with normal vector $\mathbf{N}$, gives rise to a reflected ray, $\mathbf{I}_{b1}$, and a refracted ray, $\mathbf{I}_{a2}$.

Since the slowness vector of the ray is simply $\mathbf{U} = u \cdot \mathbf{I}$, the ray vector multiplied by the scalar medium slowness, the ray/interface interactions can be represented in the 'slowness plane'. Figures 2 and 3 show this representation of the refracted and reflected rays, respectively. Since each ray vector has been multiplied by the appropriate (scalar) medium slowness, the ray angles are preserved. If the incident and refracted slowness vectors $\mathbf{U}_1$ and $\mathbf{U}_2$ are decomposed into orthogonal components ($\mathbf{P}_1$ and $\mathbf{P}_2$ parallel to the interface, $\mathbf{Q}_1$ and $\mathbf{Q}_2$ parallel to the interface normal, Figure 2), we can write

$$\mathbf{U}_2 - \mathbf{U}_1 = \mathbf{P}_2 + \mathbf{Q}_2 - \mathbf{P}_1 - \mathbf{Q}_1.$$

Snell's law requires that $\mathbf{P}_1 = \mathbf{P}_2$, so

$$\mathbf{U}_2 - \mathbf{U}_1 = \mathbf{Q}_2 - \mathbf{Q}_1.$$

The normal-parallel vectors can be obtained by

$$\mathbf{Q}_i = \mathbf{N}(\mathbf{U}_i \cdot \mathbf{N}),$$

so

$$\mathbf{U}_2 - \mathbf{U}_1 = \mathbf{N}(\mathbf{U}_2 \cdot \mathbf{N} - \mathbf{U}_1 \cdot \mathbf{N}), \tag{2a}$$

or, in terms of the ray vectors

$$u_2\mathbf{I}_2 - u_1\mathbf{I}_1 = \mathbf{N}(u_2 \cos i_2 - u_1 \cos i_1). \tag{2b}$$

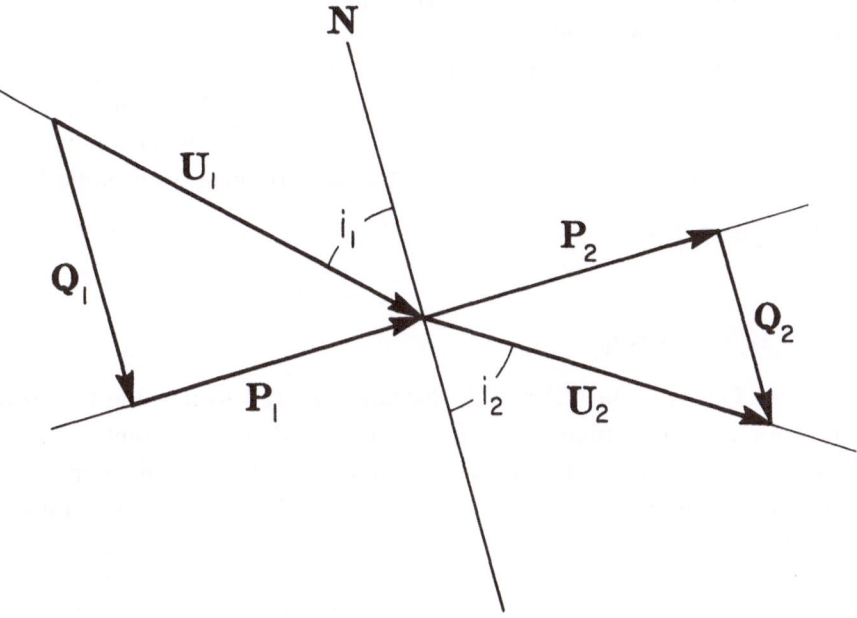

Fig. 2.   A ray refracts in the plane defined by the incident ray and the interface normal vector, $\mathbf{N}$. The rays on either side of the interface have been multiplied by the appropriate medium slowness; producing slowness vectors $\mathbf{U}_1 = u_1\mathbf{I}_1$ and $\mathbf{U}_2 = u_2\mathbf{I}_2$. These are decomposed into orthogonal vectors, $\mathbf{Q}_i$ parallel to the interface normal vector, and $\mathbf{P}_i$, parallel to the interface.

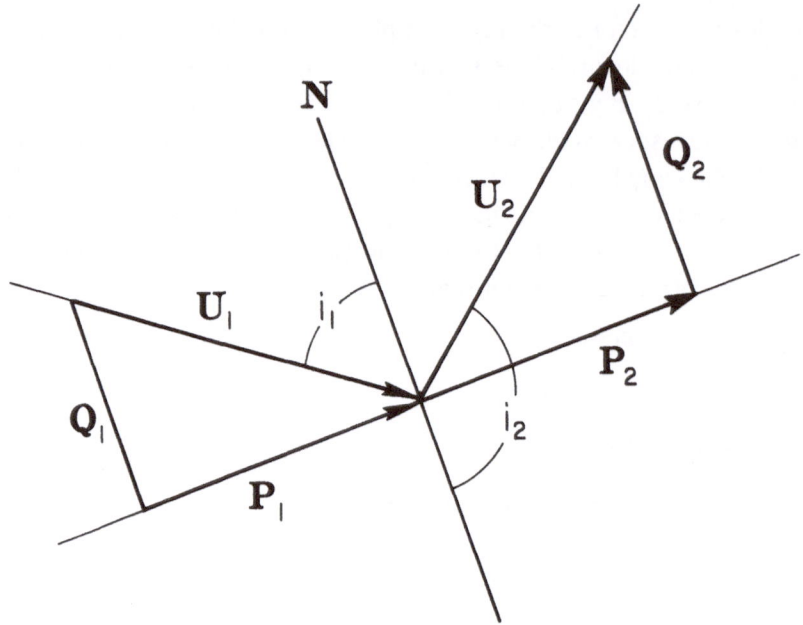

Fig. 3. A reflecting ray, shown in the plane defined by the incident ray and the interface normal. As in Figure 2, slowness vectors are shown; the result of scalar multiplication of the ray vectors and the media slownesses. These representations lead to simple derivations of the rules for ray propagation at ray/interface interactions, given by Equations (2) and (3).

A similar form is obtained for the reflected ray (Figure 3), especially in the general case, in which mode conversion takes place upon reflection, so that the incident and reflected rays have different slownesses. As before, the slowness vectors are decomposed into orthogonal elements, and Equations (2a) and (2b) are obtained again. In the special case that no mode conversion takes place, $\mathbf{I}_1 \cdot \mathbf{N} = -\mathbf{I}_2 \cdot \mathbf{N}$, and $\cos i_1 = -\cos i_2$, so the ray propagation formula becomes

$$\mathbf{U}_1 - \mathbf{U}_2 = 2\,\mathbf{N}(\mathbf{U}_1 \cdot \mathbf{N}), \tag{3a}$$

or

$$\mathbf{I}_1 - \mathbf{I}_2 = 2\mathbf{N} \cos i_1. \tag{3b}$$

Equations (1), (2), and (3) are all that are required to develop the traveltime equation for rays in homogeneous layers separated by three-dimensionally dipping interfaces. Therefore, Equations (2) and (3) also have applications in 3-D ray tracing, and are the basis for inversion of 2-D and 3-D $\tau(p)$ times to obtain layer slownesses and interface orientations.

CONTRIBUTIONS TO THE TRAVELTIME EQUATION

Consider a straight ray segment propagating from the point (X_1, Y_1, Z_1) to another point, (X_2, Y_2, Z_2) in a homogeneous medium. The traveltime for the segment can be written directly in terms of the coordinate system of the two points

$$T = p_{x2}X_2 + p_{y2}Y_2 - p_{x1}X_1 - p_{y1}Y_1 + q(Z_2 - Z_1). \tag{4}$$

Since p_{x1}, p_{y1}, p_{x2}, and p_{y2}, the horizontal slownesses of the ray segment are identical at both ends, this equation is unnecessarily complex. It is, however, of the same form as the expressions we derive below, for ray segments undergoing refraction or reflection at an interface. In all cases, the ray segment is defined by its passage between two (in the case of refractions or transits as described above), or passage back to (for reflections), horizontal planes whose Z values are the same as those of the endpoints of the ray segment.

When these three basic types of ray/interface interaction have been isolated by these horizontal planes, the resulting traveltime expressions can be simply added up to describe the traveltimes of rays whose paths can be constructed by repeated use of the ray segment 'building blocks'.

In Figure 4, the distances covered by the straight segments of ray are shown as

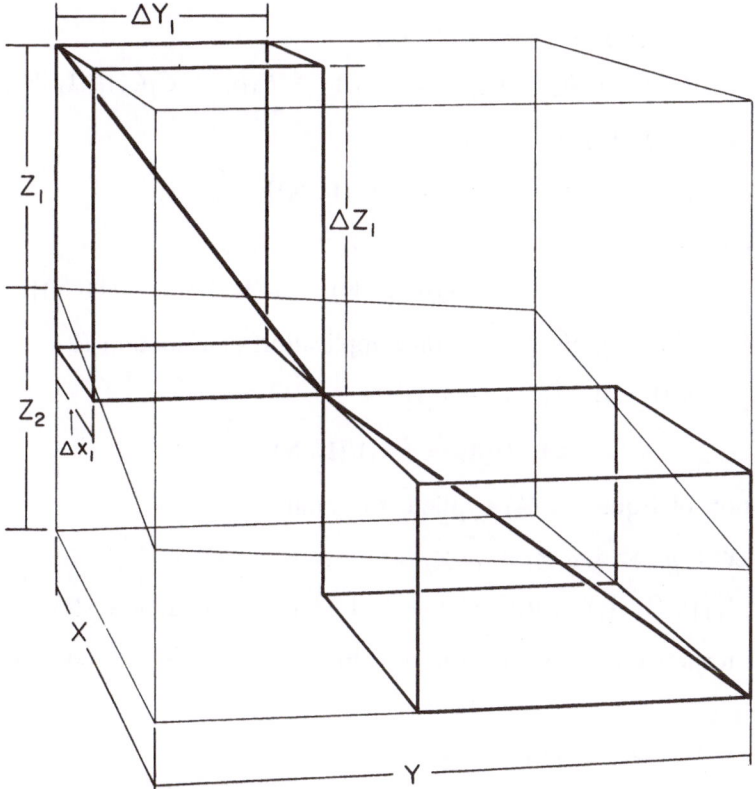

Fig. 4. The component distances travelled by the refracted ray of Figure 1 in the upper layer is ΔX_1, ΔY_1, ΔZ_1. The total distances the ray travels between the two horizontal planes are X, Y, and $Z_1 + Z_2$, the sum of the layer thicknesses measured at the ray origin. The distances traversed by the refracted ray are, then, $X - \Delta X_1$, $Y - \Delta Y_1$, $Z_1 + Z_2$.

it passes from one horizontal plane to another, refracting at an arbitrarily dipping interface along the way. The orthogonal distances covered by the ray are X, Y, and $Z_1 + Z_2$, these last being the vertical distances from the horizontal reference planes to the interface, measured at the X, Y, Z coordinate system origin. To simplify the mathematics, we choose, for the moment, the upper end of I_1 as the origin.

If the 'incident' segment $I_1 = \langle A_1, B_1, C_1 \rangle$ covers the distance ΔX_1, ΔY_1, and ΔZ_1, in the layer with slowness u_1, then the 'transmitted' segment $I_2 = \langle A_2, B_2, C_2 \rangle$ travels $X - \Delta X_1$, $Y - \Delta Y_1$, $Z_1 + Z_2 - \Delta Z_1$ with slowness u_2.

According to Equation (1),

$$T = u_1(A_1\Delta X_1 + B_1\Delta Y_1 + C_1\Delta Z_1) +$$
$$+ u_2[A_2(X-\Delta X_2) + B_2(Y-\Delta Y_2) + C_2(Z-\Delta Z_2)].$$

If the interface normal is $N = \langle \alpha, \beta, \gamma \rangle$, the point of intersection for the ray and the interface is

$$(\Delta X_1, \Delta Y_1, \Delta Z_1) = Z_1\gamma(A_1, B_1, C_1) / (I_1 \cdot N). \tag{5}$$

Substituting accordingly for ΔX_1, ΔY_1, and ΔZ_1, the traveltime becomes

$$T = u_2(A_2X + B_2Y + C_2Z_2 + C_2Z_1) +$$
$$+ Z_1\gamma[u_1(A_1^2 + B_1^2 + C_1^2) - u_2(A_1A_2 + B_1B_2 + C_1C_2)]/(I_1 \cdot N).$$

According to Equation (2), substitute

$$u_2C_2Z_1 = u_1C_1Z_1 - Z_1\gamma[u_1(I_1 \cdot N) - u_2(I_2 \cdot N)];$$
$$T = u_2(A_2X + B_2Y + C_2Z_2) + u_1C_1Z_1 +$$
$$+ Z_1\gamma\{[u_1(I_1 \cdot I_1) - u_2(I_1 \cdot I_2)]/(I_1 \cdot N) - u_1(I_1 \cdot N) + u_2(I_2 \cdot N)\}.$$

The last term can be simplified by further application of Equation (2)

$$u_2(I_1 \cdot N) - u_1(I_1 \cdot N) = (u_2I_2 - u_1I_1)/N$$
$$= I_1 \cdot (u_2I_2 - u_1I_1)/(I_1 \cdot N),$$

and the notation of Equation (4) applied, to obtain

$$T = p_{x2}X + p_{y2}Y + q_1Z_1 + q_2Z_2 +$$
$$+ Z_1\gamma[u_1(I_1 \cdot I_1) - u_2(I_1 \cdot I_2) + u_2(I_1 \cdot I_2) - u_1(I_1 \cdot I_1)]/(I_1 \cdot N).$$

Since the last term is equal to zero, the traveltime for this representative refracting ray is

$$T = p_{x2}X + p_{y2}Y + q_1Z_1 + q_2Z_2.$$

Although the coordinate origin in this development was chosen for convenience, the basic form of the equation will be the same for any choice. If the origin is shifted horizontally, additional terms will arise

$$T = p_{x1}X_1 + p_{x2}X_2 + p_{y1}Y_1 + p_{y2}Y_2 + q_1Z_1 + q_2Z_2. \tag{6}$$

Equation (6) is similar to Equation (4); the horizontal distances to the ends of the rays X_1, Y_1, X_2, and Y_2 are multiplied by the appropriate horizontal slownesses, and the layers' thicknesses Z_1, Z_2, as measured vertically from the coordinate origin, are multiplied by the vertical ray slownesses.

The third and final element required is an expression for the traveltime for a ray reflecting from an arbitrarily dipping interface. Figure 5 shows such a ray 'originating' from a horizontal reference plane, reflecting from the interface with normal vector $\mathbf{N}$ and returning to the original level. As in Figure 4, the distances covered by the incident ray are ΔX_1, ΔY_1, and ΔZ_1, but since the reflecting ray returns to the level of its origin, it traverses $X-\Delta X_1$, $Y-\Delta Y_1$, and $-\Delta Z_1$. If $\mathbf{I}_1$ is the incident and $\mathbf{I}_2$ the reflected ray, ΔX_1, ΔY_1, and ΔZ_1 are again given by Equation (5), and the traveltime by

$$T/u = A_2X + B_2Y + Z_1\gamma[(\mathbf{I}_1 \cdot \mathbf{I}_1) - (\mathbf{I}_1 \cdot \mathbf{I}_2)]/(\mathbf{I}_1 \cdot \mathbf{N}).$$

This time, the last term is not equal to zero and the terms in C_1 and C_2 are

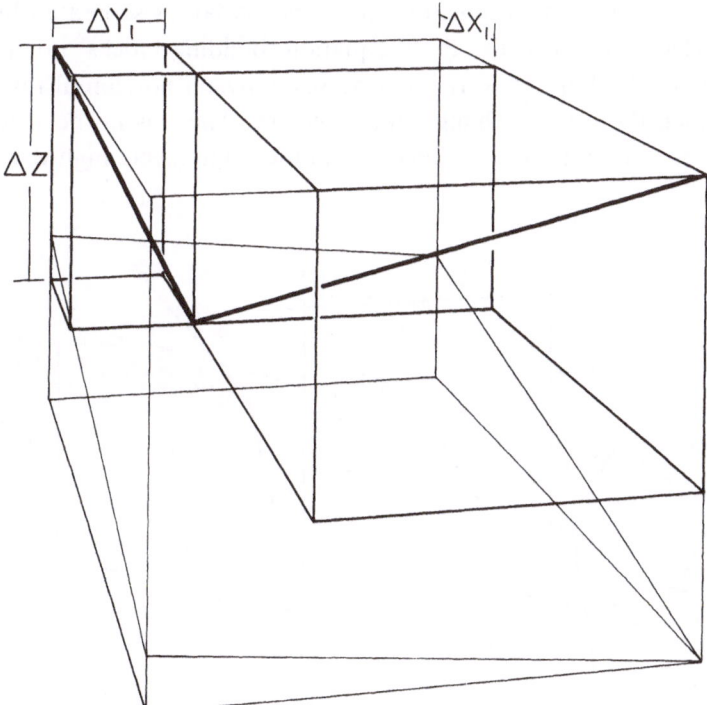

Fig. 5. When the incident ray of Figure 4 is reflected, the distances traversed by the reflected ray are $X - \Delta X_1$, $Y - \Delta Y_1$, $Z - \Delta Z$.

missing. Both of these problems can be corrected by adding some terms according to Equation (3)

$$Z_1(C_1 - C_2) + Z_1\gamma[(\mathbf{I}_2 \cdot \mathbf{N}) - (\mathbf{I}_1 \cdot \mathbf{N})] = 0,$$

which can be manipulated to yield

$$Z_1(C_1 - C_2) + Z_1\gamma \mathbf{I}_1 \cdot (\mathbf{I}_2 - \mathbf{I}_1)/(\mathbf{I}_1 \cdot \mathbf{N}) = 0.$$

This step, along with the simplifications used in the development of Equation (6), give

$$T = p_{x2} X + p_{y2}Y + q_1 Z_1 + q_2 Z_2$$

again, since the other terms cancel each other as before. This equation is essentially identical to Equation (6), though attention must be paid to the signs of the vertical slownesses and of the vertical distances travelled.

3-D TRAVELTIME EQUATION

Using the expressions for the traveltime contributions of the three elementary types of ray interaction developed above (refraction, refraction, no change), the traveltime equation can be 'assembled' for any source--receiver geometry within structures that satisfy the assumptions of homogeneous layers separated by dipping, planar interfaces. Each of the three equation 'building blocks' isolates the effects of changes in the horizontal ray slownesses between horizontal reference planes, across which there is no change in slownesses. There is no need for the source and receiver to be on the same level, or in the same layer. Figure 6 shows

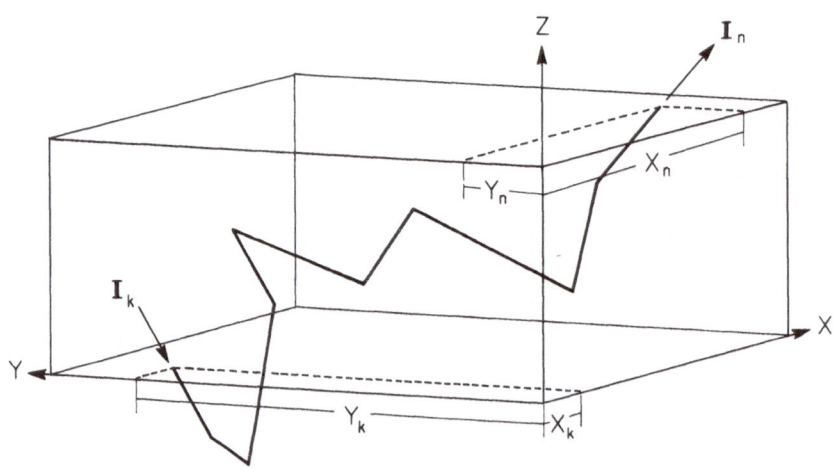

Fig. 6. An incident ray, $\mathbf{I}_k$, is refracted and reflected through a structure defined by homogeneous layers and planar interfaces. The horizontal offsets, from the arbitrary origin to the entering and exiting ray points are shown: X_k, Y_k, X_n, and Y_n. The traveltime for this ray segment is given by Equation (7)

$$T = p_{xk}X_k + p_{yk}Y_k - p_{xn}X_n - p_{yn}Y_n + \Sigma q_i Z_i.$$

a very general ray, starting at X_k, Y_k, with slowness $u_k\mathbf{I}_k$. This ray propagates through a [not shown] structure of planar interfaces, obeying Snell's law, and Equations (2) and (3). It emerges at X_n, Y_n, with slowness $u_n\mathbf{I}_n$. The ray's path may include multiples and phase conversions. The traveltime equation for any such ray is

$$T = p_{xk}X_k + P_{yk}Y_k - P_{xn}X_n - p_{yn}Y_n + \sum q_i Z_i, \tag{7}$$

where the sum is carried out so that i is equal to the index of every layer traversed.

Equation (7) is the general traveltime for rays in a structure made of homogeneous layers separated by arbitrarily dipping planar interfaces. Equation (7) is also valid for nonplanar interfaces, as long as the $\mathbf{N}$'s in Equations (2) and (3) are normals to the interface surfaces at the points of ray intersection. This generalization, however, does not facilitate Equation (7)'s use in $\tau(p)$ inversion. For practical application, the coordinate origin is chosen to minimize the number of terms in the equation and so as to correspond to a fixed point in the geometry of the profile under analysis. The orientation of the horizontal (X, Y) axes can be similarly chosen (to line up with a straight line of shots to a fixed receiver at $X = Y = 0$, for example). In cases that one-dimensional or two-dimensional structures are to be assumed in the inversion, the 'extraneous' terms in Equation (7) will serve to quantify the errors arising from violations of the assumption.

3-D INTERCEPT TIME

No matter what profile geometry is used, the accumulated vertical delay terms will have virtually identical form, corresponding to the 'correct' intercept time

$$\tau = \sum q_i Z_i \tag{8}$$

which will result if the horizontal phase delay terms are completely removed.

Velocity-Depth Inversion from One-Dimensional Intercept Times

The remainder of this chapter concerns the analysis of $\tau(p)$ data to determine the variation of seismic velocity with depth in the earth. In every case, we assume that observed $T(X)$ data have been transformed into a $\tau(p)$ data set comprising a number of traces, each giving amplitude as a function of intercept time for a particular ray parameter. Since the postcritical inversions presented here are done in a top-down manner, and because the shallowest velocities encountered in the earth usually correspond to the largest values of p that can be observed, it is convenient to number the traces, and therefore the p's, starting with $p_1 = p_{\max}$, the slowness at the surface (Figure 7).

When velocity is a function of depth alone, and when layers are assumed to be homogeneous, the appropriate forms of the discrete traveltime and intercept time

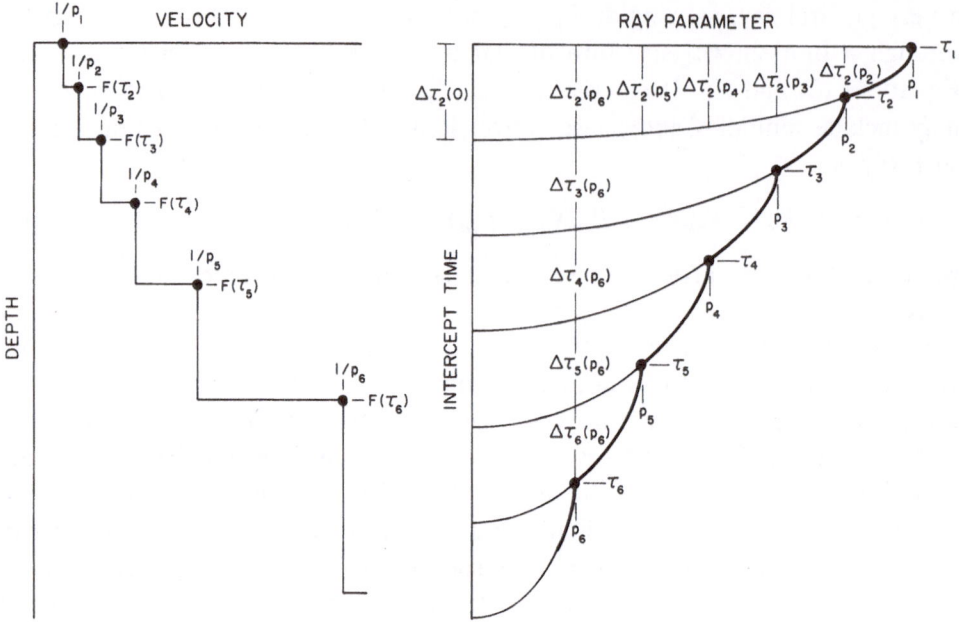

Fig. 7. The notation of Equations (10) is illustrated; each discretely sampled postcritical $\tau(p)$ point τ_n is numbered consecutively, beginning with the highest values of p, which tend to be the lowest values of τ. Sampled or predicted values of precritical $\tau(p)$ points are numbered consecutively with increasing values of τ. $\text{Tau}_1(p_j)$ is always zero.

equations are quite simple:

$$T_n(p_m) = p_m X + \tau_n(p_m),$$

where

$$\tau_n(p_m) = 2\sum_{j=1}^{n-1} Z_j(u_j^2 - p_m^2)^{1/2}$$

$$= 2\sum_{j=1}^{n-1} Z_j q_j(p_m). \tag{9}$$

Inversions of Equation (9) for velocities ($1/u_j$) and layer thicknesses (Z_j) fall into two groups; those using only postcritical arrivals and those that can employ precritical arrivals as well. All of these methods are ray theoretical in that they are based on estimations of geometrical arrival and intercept times, but since these are always related to a specific ray parameter, it is possible to include angle dependent effects, if desired.

The basic equation used in all of the inversion methods considered here can be derived from the discrete τ equation (9);

$$\Delta\tau_n(p_m) = 2Z_{n-1} q_{n-1}(p_m), \tag{10a}$$

where

$$\Delta\tau_n(p_m) = \tau_n(p_m) - \tau_{n-1}(p_m), \tag{10b}$$

and

$$q_{n-1}(p_m) = (p_{n-1}^2 - p_m^2)^{1/2}. \tag{10c}$$

To simplify notation, $\tau_n(p_n)$, a sampled, postcritical point will be written τ_n. The indexing scheme for Equations (10) and the rest of this chapter is illustrated in Figure 7.

1-D POSTCRITICAL ARRIVAL INVERSION

One of the first features of the $\tau(p)$ transformation to be noticed was that it separates postcritical and precritical arrivals. If we consider only a single mode of phase propagation, the postcritical arrivals will form a trajectory with intercept times increasing as ray parameter decreases (Figure 8). This postcritical path will be broken only when data gaps, or shadow zones due to low velocity layers, are present. Since this postcritical energy can be easily 'picked', by virtue of its high amplitude and location, velocity-depth functions can be repidly obtained by the postcritical methods. Perhaps the simplest form of inversion for postcritical arrivals is the 'Tau-sum' method of Diebold and Stoffa (1981); Diebold *et al.* (1981). This method entails the recursive solution of the discrete intercept time equation (9). In the original and simplest form of the method, an homogeneous layer is considered to exist for each picked $\tau(p)$ pair. The slowness of any layer is

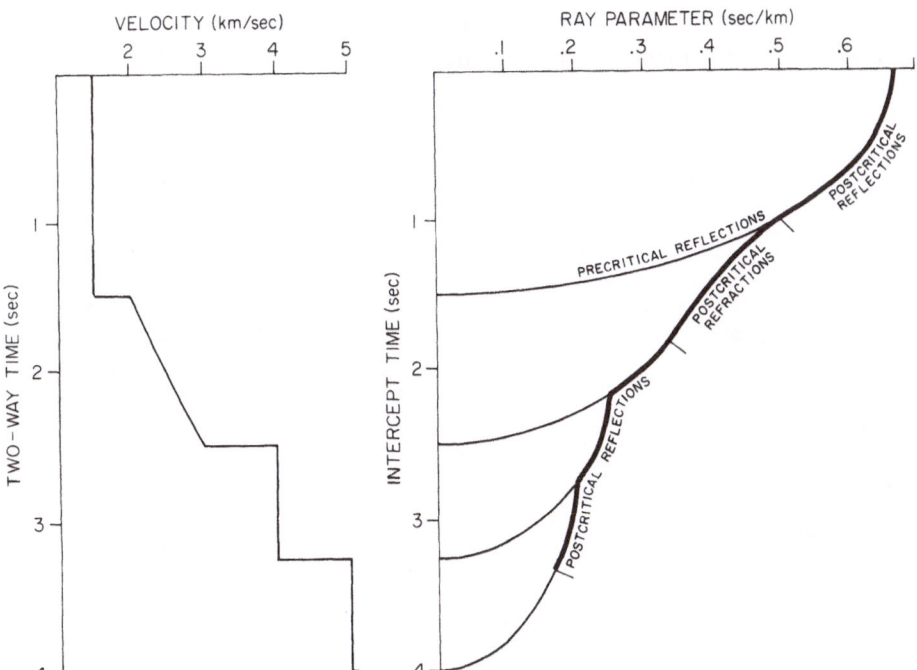

Fig. 8. A simple, one-dimensional velocity model (left) and the corresponding $\tau(p)$ plot (right). The $\tau(p)$ transformation separates the precritical and postcritical arrivals, as indicated.

given by the corresponding ray parameter $u_j = p_j$, and the layer thickness is given by Equation (10a), when $m = n$,

$$Z_{n-1} = \Delta\tau_n(p_n)/[2q_{n-1}(p_n)]. \tag{11}$$

Each $\Delta\tau$ is determined by a postcritical pick, τ_n, and $\tau_{n-1}(p_n)$, which is calculated based on previous solutions for the overlying Z's. The necessary top-down progress of this procedure is responsible for the numbering of the picked $\tau(p)$ points starting with the largest values of p, as indicated in Figure 7. Figure 9 shows a good example of Tau-sum inversion applied to data from a wide aperture common midpoint survey (a two-ship expanding spread profile, or 'ESP').

This layer-by-layer downwards 'bootstrap' procedure requires that the solution begins at the earth's surface (or at whatever depth the sources and receivers may lie). A problem arises when the first postcritical arrivals seen in the transformed data have been returned from a deeper layer. Usually, the velocity of the 'surface' layer is known or can be obtained from the data by precritical methods. Based on this information, a dummy point, $p_1 = 1./V_{\text{surface}}$, $\tau_1 = 0.$, is included in the data set. As a result, the Tau-sum recursion will produce a first layer with the proper velocity, and whose thickness will be defined by the location of τ_2.

When, as in the case of marine data, the 'surface' is in or on a more or less homogeneous layer, the intercept time of the first *sampled* postcritical $\tau(p)$ point will not usually give the correct thickness for the first layer and a second dummy point must be included in the data set. When the thickness of the surficial layer and the velocity of the layer just below are known, this point can be easily calculated from Equation (9). Often, however, although the thickness of the surficial layer is known, the next velocity is not. In this case, it must be guessed

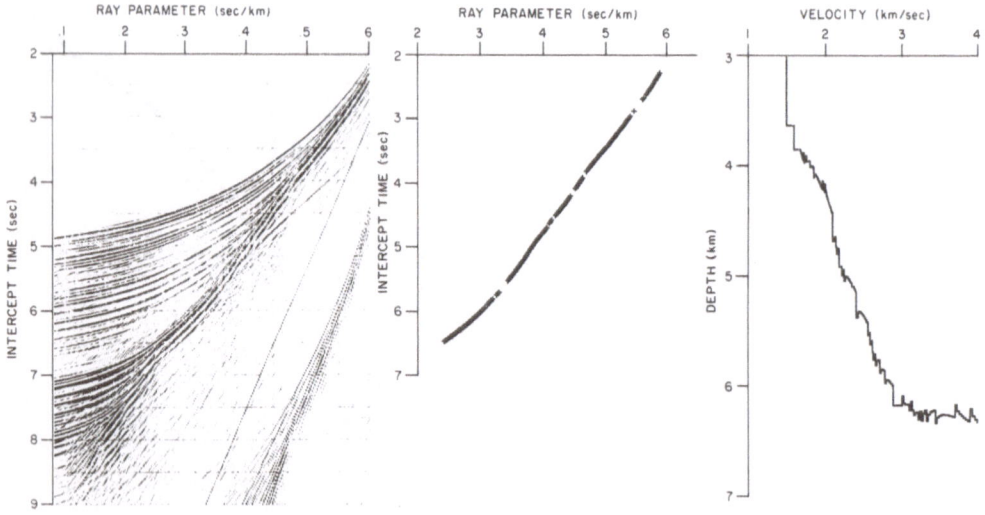

Fig. 9.   Tau-sum inversion results (right), based on 150 postcritical $\tau(p)$ picks (center) of a typical data set (left).

and (9) used to calculate τ_2 for the corresponding p_2. If too high a velocity is used, the thickness of the second layer will be overestimated by the recursion, and a negative thickness will be calculated for the third layer. Some simple analysis, however, will suffice to define the range of permissible points. The maximum allowable velocity for the second layer is that producing $Z_3 = 0.$, and therefore, $\Delta\tau_4(p_4) = 0$. In this case, Equation (9) can be written for the first two sampled $\tau(p)$ points:

$$\tau_3 = \Delta\tau_2(p_3) + \Delta\tau_3(p_3) \quad \text{and} \quad \tau_4 = \Delta\tau_2(p_4) + \Delta\tau_3(p_4).$$

Separating terms in Z_2 yields

$$\Delta\tau_3(p_3) = 2Z_2 q_2(p_3) = \tau_3 - \Delta\tau_2(p_3),$$

and

$$\Delta\tau_3(p_4) = 2Z_2 q_2(p_4) = \Delta\tau_4 - \Delta\tau_2(p_4).$$

Eliminating $2Z_2$,

$$[\tau_3 - \Delta\tau_2(p_3)]/q_2(p_3) = [\tau_4 - \Delta\tau_2(p_4)]/q_2(p_4)$$

is obtained. By substituting for the q_2's according to (10c), squaring out, and separating terms in p_2^2, the minimum value

$$p_2^2 = \frac{p_3^2[\tau_4 - \Delta\tau_2(p_4)]^2 - p_4^2[\tau_3 - \Delta\tau_2(p_3)]^2}{[\tau_4 - \Delta\tau_2(p_4)]^2 - [\tau_3 - \Delta\tau_2(p_3)]^2}$$

can be easily determined.

As Diebold and Stoffa (1981) pointed out, the Tau-sum recursion is mathematically identical to the 'slope-intercept' method classically used to derive $V(Z)$ functions from headwave picks (Slotnick, 1936), but they showed that it is equally well suited to the interpretation of postcritical arrivals of any kind. When several $\tau(p)$ points along a postcritical reflection are considered in sequence, the first gives the thickness of the layer above, and a velocity somewhere between that of the layers on either side of the discontinuity. Subsequent recursions for the other $\tau(p)$ points result in determinations of zero layer thicknesses and velocities that are increasingly closer to that of the layer below.

Another familiar method of velocity-depth inversion from postcritical arrivals is the Herglotz–Weichert–Bateman [HWB] integral method, which was originally developed for use in the $X(p)$ domain. Several authors have implemented this technique in the $\tau(p)$ domain (Bessonova et al., 1974; Kennett, 1976). Since HWB is an integral method, its application to discrete data requires a point-to-point interpolation. Recently, Vera (1987) has shown that when the HWB interpolation paths in $\tau(p)$ are the pseudo-ellipses corresponding to homogeneous layers, the HWB and Tau-sum methods are mathematically identical. Therefore, the Tau-sum recursion is not necessarily 'less linear' than any inversion method based on the HWB integral.

A great amount of development in $\tau(p)$ analysis has been motivated by the difficulties and uncertainties inherent in the determination of $\tau(p)$ values from worldwide earthquake traveltime observations. Johnson and Lee (1985), for example, used some 60 000 observations to determine 31 $\tau(p)$ points with bounds between ±48 msec and ±2403 msec, which they use to estimate extremal bounds for P velocity over 3500 km of the Earth's core. Other workers, Stark et al. (1986) for example, have applied various implementations of the linear programming method to the same ends. Since the exploration data discussed in this chapter are typically sufficient to produce as many as 200 directly determined $\tau(p)$ points, which sample a few tens of kilometers of the Earth's crust (that is, having a sample density at least three orders of magnitude over that used in the extremal inversion studies cited above), we have chosen to stick to the simplistic, but exact solutions described here.

The discrete τ equation (9) represents an expansion of the continuous intercept time integral, $\tau(p) = \int q(Z, p) \, dZ$, in terms of homogeneous layers. The equation of the seismic slownesses in the final model with the sampled p's, results in a set of linear equations that can be solved by an equivalent number of layers. A similar set of discrete equations can be written using as a basis *any* velocity function for which an analytic solution for τ can be found. Some of these functions are listed by Kaufman (1953), Garmany et al. (1979), and Greenhalgh and King, (1981). The Tau-sum recursion can be modified so as to use any of these analytic functions of τ as a kernel (Vera and Diebold, 1984). Once an analytic $\tau(p)$ function has been chosen, the appropriate interpolation path is automatically at hand if HWB inversion is preferred. As Diebold and Stoffa (1981) show, however, Tau-sum inversion based on the homogeneous layer expansion is quite adequate to approximate more complicated velocity functions when the $\tau(p)$ field is sampled finely enough (see, for example, Figures 13 and 15, below). The converse is not always true. Tau-sum recursion with a velocity gradient kernel is quite stable for data from velocity gradients, but can be unstable when applied to large-offset postcritical reflections from homogeneous layers.

1-D PRECRITICAL ARRIVAL INVERSION

Although the analysis of postcritical arrivals can provide detailed information about complicated velocity functions, uncertainties can arise when the postcritical $\tau(p)$ trajectory is broken, a situation that can result from the presence of velocity reversals, or from limitations in offset coverage. In these cases, *ad-hoc* adjustments must be made to preserve, as best as possible, the accuracy of the results for deeper layers. Inversions based on the analysis of precritical reflections can still be successfully performed, however, without discrimination. In standard CDP data, postcritical methods are applicable for only the shallowest part of the section, if at all, and inversions based on precritical arrivals are more appropriate.

When precritical arrivals are used for velocity-depth inversion, (10a) is still the controlling equation. Since the postcritical arrival with the layer slowness cannot be assumed to have been directly sampled, at least two observations of $\Delta\tau(p)$ must be made to solve a single layer's slowness, as well as its thickness (Schultz, 1976). The resolution of the inversion can be improved when $\Delta\tau(p)$ is determined over a large range of ray parameters. Equation (10a) describes an ellipse, with semi-major and minor axes of length $\Delta\tau_n(0)$ and p_{n-1}, the interval two-way time and slowness of the layer. These parameters can be extracted from a linear least-squares fit to squared $\Delta\tau_n(p)$ values, plotted against p^2 (Figure 10). This process is mathematically identical to the layer stripping/elliptical semblance search method described by Schultz (1982) and Schultz $et\ al.$ (1983). An elliptical $\tau(p)$ path is, of course, equivalent to a $T(X)$ hyperbola, but this does not imply that $\Delta\tau^2 - p^2$ inversion is equivalent to $T^2 - X^2$ techniques, except in the case that each precritical $\tau(p)$ reflection branch is simply fitted by a single ellipse (Cutler and Love, 1980). The great advantage of the $\tau(p)$ transformation is that the $\Delta\tau_n(p_m)$ values can be easily extracted, and interval slowness (and, equivalently, velocity) can be determined directly and without the well-known offset bias inherent in $T^2 - X^2$ methods.

Extraction of the $\Delta\tau(p)$ values can be done by a variety of means. Schultz (1982) describes an interactive layer stripping method; a top-down procedure of picking the minimum $\Delta\tau(0) - p$ peak in an elliptical semblance plot, moving

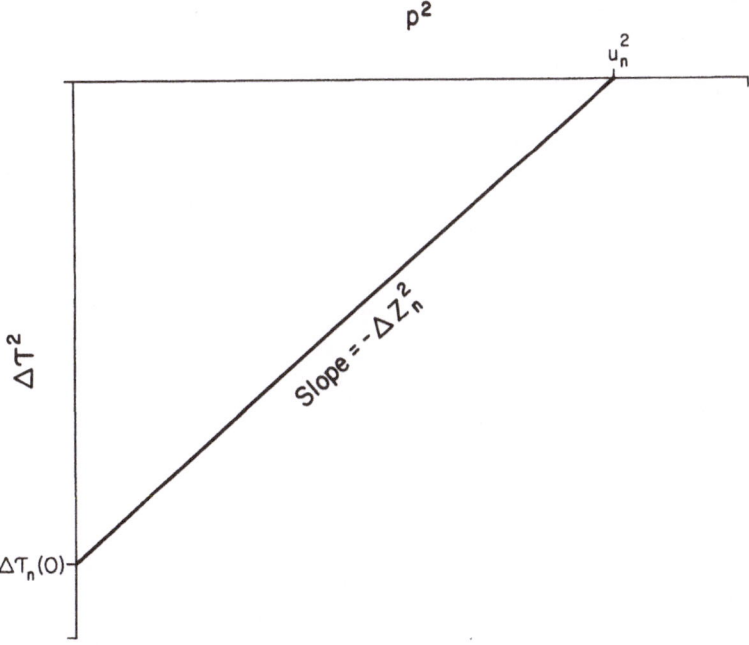

Fig. 10. When the elliptical $\Delta\tau(p)$ function of Equation (10) is squared and plotted against p^2, the result is a straight line, whose slope and intercepts can be interpreted to give the thickness and slowness of a homogeneous layer.

out the $\tau(p)$ data to remove the effect of the corresponding ellipse, making a
new semblance search, and picking again. Due to the moveout step, the shape of
each new ellipse is affected by the previous results, so that the inversion of each
layer is not completely independent of the others, as in the case of $T^2 - X^2$
inversion.

The example in Figure 11 was produced by a simpler method, which is somewhat less computationally intensive. Tau (p) reflection points were obtained by

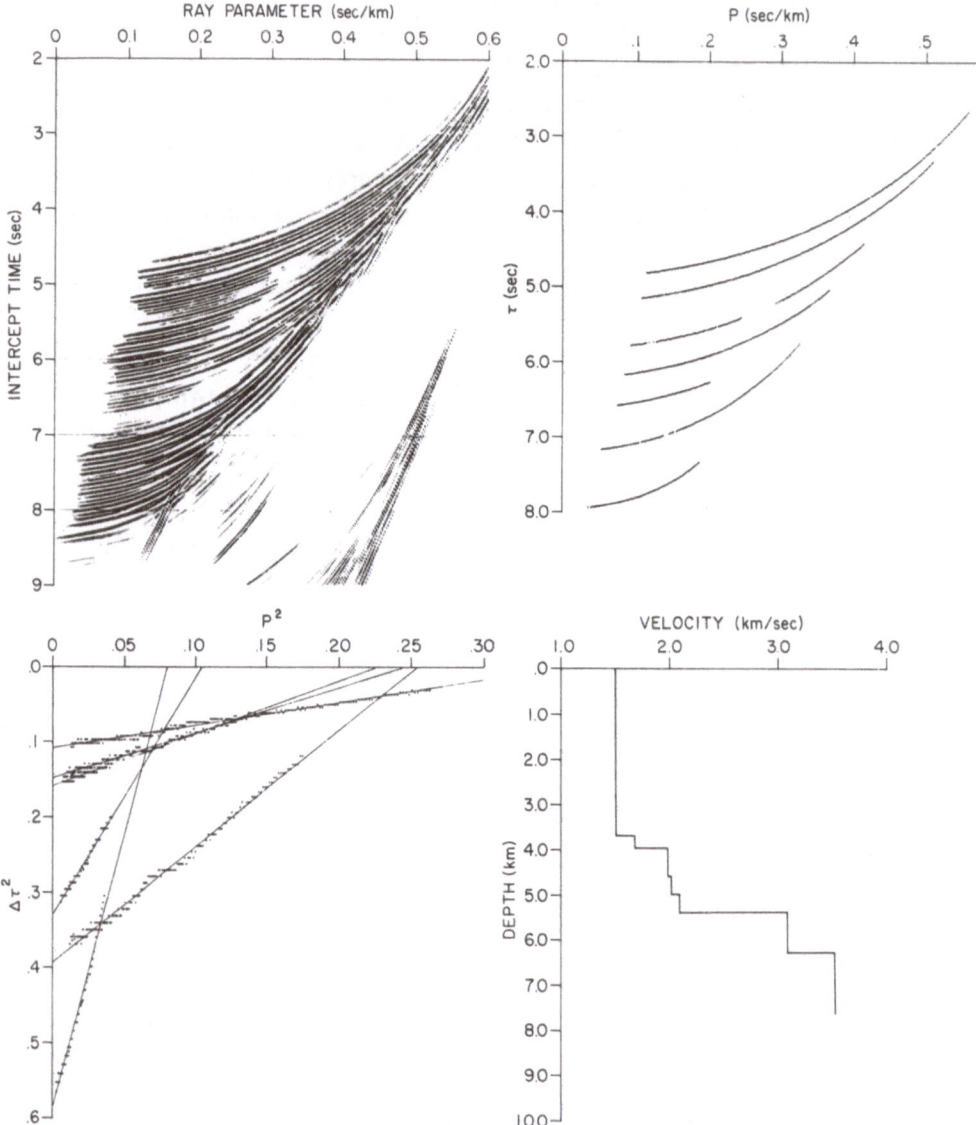

Fig. 11. $\Delta\tau^2-p^2$ inversion example. Reflections in $\tau(p)$ data (a) are picked (b). From these,
$\Delta\tau^2-p^2$ points are derived, and fitted by straight lines (c). The resulting $V - Z$ pairs are combined
to produce the velocity-depth model (d).

automatic, but guided picking. Each reflection interval was treated in turn, in typical top-down fashion. $\Delta\tau_n(p)$ was calculated for each pair of picks. Layer thickness and slowness was then obtained by simple least-squares fitting of the $\Delta\tau^2 - p^2$ values. Errors in this method do not propagate downwards, until the individual layer solutions are superimposed to obtain the final solution (Figure 11(d)).

One of the great advantages of working with data in the $\tau(p)$ domain is that most, if not all, of the well-known $T(X)$ methods can be applied in $\tau(p)$, though the converse is not true. An example is the elliptical velocity search (Cutler and Love, 1980), which can be performed by making coherency stacks along straight lines in $\tau^2 - p^2$ space, without the layer stripping described above. This process will produce a two-dimensional velocity *vs.* two-way time coherency plot, such as that resulting from standard hyperbolic $T(X)$ velocity searches. The velocity function picked from these data can then be used to move out the $\tau(p)$ data and stack it (Stoffa *et al.*, 1982) and interval velocities can be determined using Dix's (1955) equation.

LIMITATIONS

Both of the inversion methods outlined above have limitations. In many cases, the limitations can be overcome by combining the two methods. In any case, it is important to understand where the methods' relative accuracies break down, and to refrain from applying them blindly in inappropriate circumstances.

TAU-SUM ERROR RESPONSE

The sensitivity of the Tau-sum equation (11) to errors in $\Delta\tau$ depends on the inverse of $q_{n-1}(p_n)$. There is a tradeoff between the slowness resolution implied by the p sampling interval and the resolution of thickness, which is a function of the resolution with which $\Delta\tau$ is known, and Equation (11). When $\delta p = p_{n-1} - p_n$ is small, $q_{n-1}(p_n)_2 \simeq 2\delta p\, p_n$, so the layer thickness error has a constant $(8\delta p)^{-1/2}$ term, and a $p^{-1/2}$ dependence. A typical value, for a p interval of 0.002 sec/km and a ray parameter of 0.5, is 11 m of Z error for every 1 msec of $\Delta\tau$ error.

Fortunately, errors introduced into the Tau-sum inversion tend to be self correcting, rather than producing a cumulative effect. The inversion is recursive, in that solutions for Z_j are used to calculate $\Delta\tau$ for deeper layers, according to Equation (10). An initial positive error in Z_j will produce a larger negative error in Z_{j+1}, a smaller positive error in Z_{j+2}, and smaller positive errors in succeeding recursions. This process is illustrated in Figure 12. Unless the initial error is very large, the cumulative effect will be greatly diminished after a few subsequent recursions.

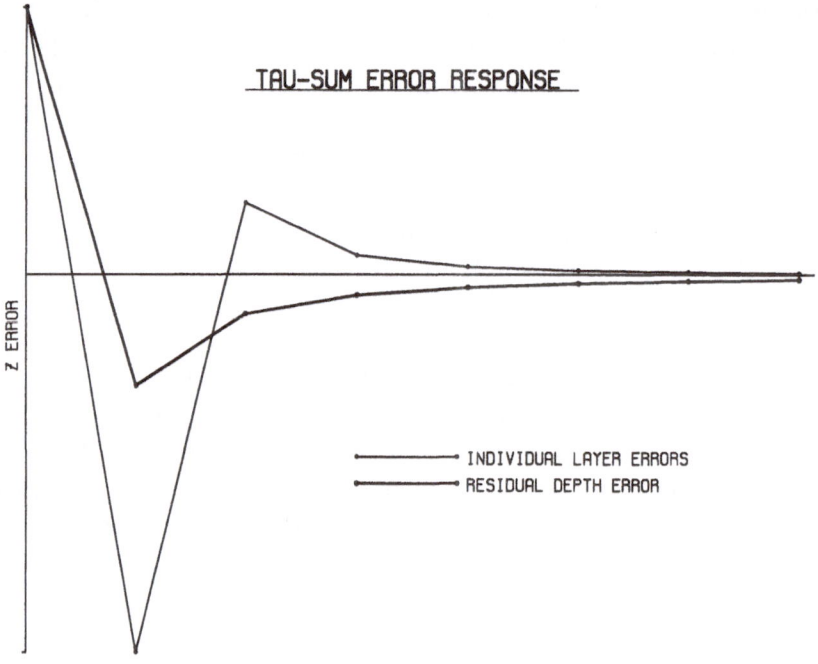

rïg. 12.   The error response of the Tau-sum recursion. When a positive $\Delta\tau_n$ error is introduced, it produces an initial positive error in Z as calculated by the recursion. The error propagates during the recursion for deeper layers, producing additional errors of alternating sign (thin line). Since the propagated error has oscillating terms, the cumulative effect (thick line) is self-correcting.

INTERCEPT TIME UNCERTAINTY

There are many possible sources of τ uncertainty, including bandwidth limitations of source and receiver systems, frequency-dependent attenuation in the earth, and phase shifting of reflected and refracted arrivals. Some of these can be overcome by improved experimental and processing methods, and others by more sophisticated inversion methods. The fundamental limit on τ resolution, however, is the $\tau(p)$ seismogram time sample interval. In Figure 13, Tau-sum inversion has been applied to raytraced intercept times for a simple model, which have been truncated to simulate a sample rate of 4 msec; a typical value. As predicted by Figure 12, the deviations are not cumulative, but they increase in amplitude with increasing velocity. This effect is commonly seen in Tau-sum inversions of real data. Figure 13(b) compares part of this synthetic result with part of the actual inversion of Figure 9. Although the model for Figure 13 was not intended to resemble the inversion in Figure 9, they each have a velocity discontinuity between 3.5 and 4 km/sec, which produces postcritical $\tau(p)$ reflections with similar slopes. The result is that the sequence of errors due to the time sampling is nearly identical.

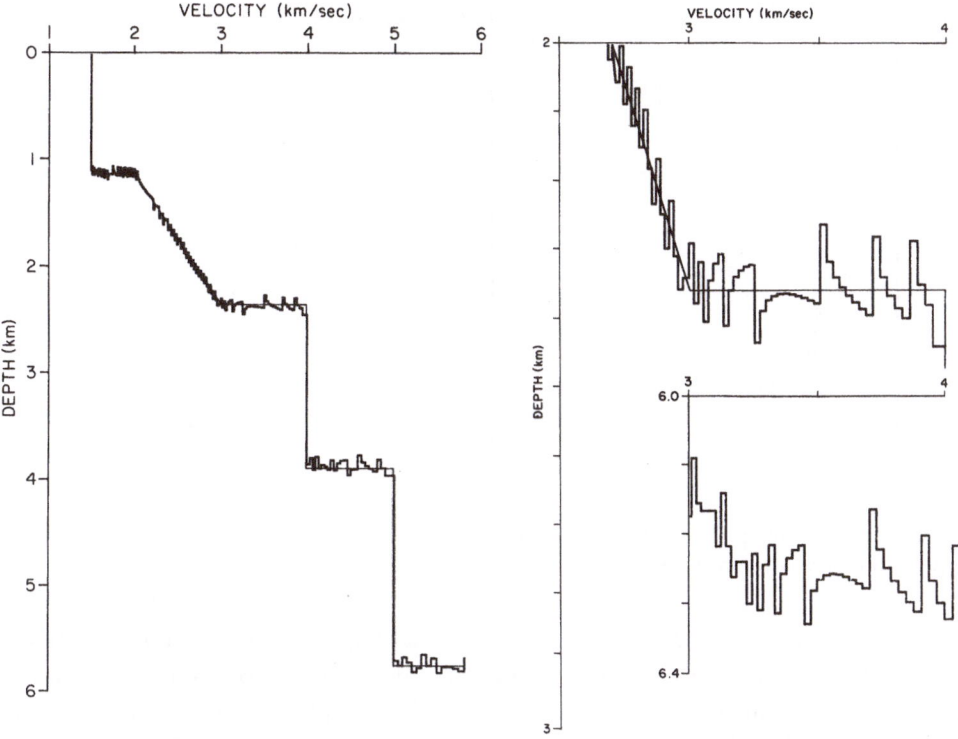

Fig. 13(a). The effect of a finite time sampling interval on the Tau-sum inversion of otherwise perfect data. Raytraced intercept times for a simple model (smooth line), when truncated to simulate a 250 Hz sampling rate, with a ray parameter spacing dp = 0.002 sec/km, produce a Tau-sum inversion with appreciable jitter.

Fig. 13(b). Similar effects are seen in the Tau-sum inversion of real data. A portion of the synthetic result from Figure 13(a) (top) is compared to a part of the inversion result of Figure 9 (bottom), at the same scale.

Three approaches are possible in dealing with the jitter caused by finite time sampling. If the data bandwidth justifies it, the sample rate can be made smaller by resampling the $T(X)$ traces before the $\tau(p)$ transformation. A smoothing filter can be run across the $\tau(p)$ picks, with an implicit degradation of p (and V) resolution, or the sample error envelope, which can be easily calculated, can be simply accepted as the resolution limit.

POSTCRITICAL ARRIVAL DATA GAPS

Tau-sum inversion may produce layers with grossly negative thicknesses. From Equation (11), it is seen that a negative Z_n results directly from a negative $\Delta\tau_{n+1}$ (Figure 14). Neglecting the effects of sample jitter and topography, a negative $\Delta\tau$ usually results from one or both of two causes; missing postcritical

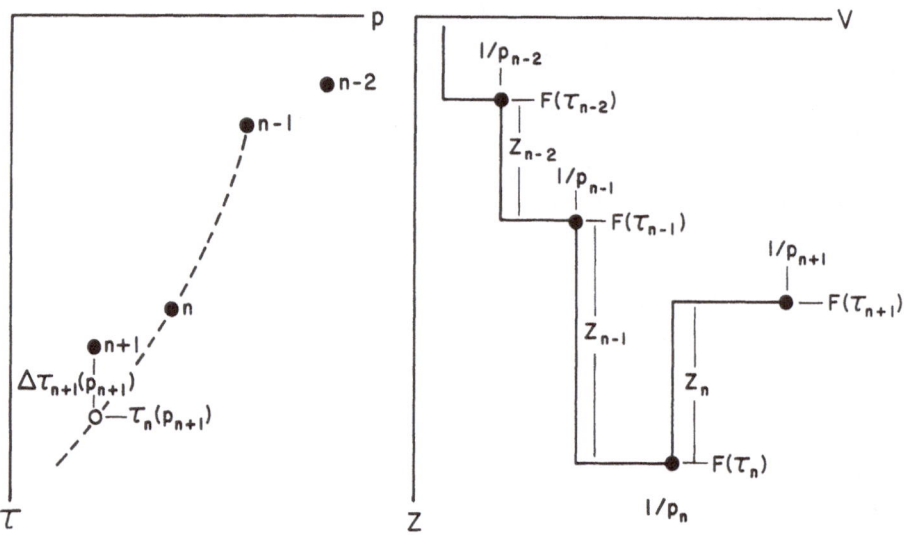

Fig. 14. A sharp jump in sampled $\tau(p)$ values can produce a negative $\Delta\tau(p)$ (left), producing a negative thickness layer in the Tau-sum inversion (right).

data (τ_{n-1} is precritical), or that $\tau_{n-1} - \tau_n$ is a gap in the postcritical arrival path, caused by an LVZ. A serious, and well known limitation of any postcritical arrival inversion method is its inability to directly produce the velocities and thicknesses of LVZ's. In the case of the Tau-sum inversion, the lack of a post-critical pick for an LVZ actually causes the thickness of the layer immediately overlying the LVZ and the LVZ's slowness to be undefined. A similar problem is met where postcritical data are missing, as a result of structure, or of limitations in offset coverage (Brocher and Phinney, 1981). In either case, limits can be placed on the missing information by analysis of the requirement that

$$\Delta\tau_{n+1}(p_{n+1}) \geqslant 0.$$

Performing the correction for missing postcritical data is quite simple, and exactly correct, as long as the missing points are just the maximum offset contributions to the postcritical reflections from the interface beneath a single homogeneous layer. In this case, the situation can be corrected exactly by finding the last good (i.e., postcritical) sample. In the case shown in Figure 14, τ_{n-1} obviously won't do. The procedure is to work backwards, starting with $k = n-2$, to find the highest k such that

$$\tau_{n+1} \geqslant 2 \sum_{i=1}^{k} Z_i q_i(p_{n+1}).$$

The intervening (precritical) points are then eliminated.

Gaps in postcritical data coverage are most likely to arise from two sources; insufficient offset in the $T(X)$ data set, and topographic effects. In order to detect the latter, it is a good idea to examine near vertical reflection data recorded along the profile. If topographic effects along any interface are large enough to cause postcritical data gaps, the $T(X)$ data are not appropriate for slant stacking in the first place, unless prestack migration or layer replacement is also performed. In the absence of variable interface topography, the effect of limited offset may, in some cases, be signalled by the fact that the $\tau(p)$ slope on the low-p side of the gap is incompatible with the maximum offset in the $T(X)$ data. Another criterion, which was the basis of Brocher and Phinney's (1981) method, is arrival amplitude. Ideally, the erroneously included precritical arrivals will also be distinguishable from the postcritical events by their smaller amplitudes.

Given knowledge of the velocities and thicknesses of LVZ's, it is a trivial problem to include them in the Tau-sum process, as long as they can be represented by one or more homogeneous layers. This can be done by using the discrete forward τ equation (9) to create a synthetic $\tau(p)$ point $[\tau_{\mathrm{LVZ}}(p_{\mathrm{LVZ}})]$ for each LVZ. In the context of Tau-sum recursion, this missing point will provide the velocity in the LVZ, and the thickness of the layer above, sometimes called the 'high velocity cap' (HVC). The thickness of the LVZ itself is given by the next intercept time. When these synthetic points are in place, the absolute value of the argument inside the square root of the q term, as given in Equation (10c) must be used in both the forward and inverse steps, since the 'critical' q for an LVZ is an imaginary quantity.

An infinity of synthetic points could be chosen, corresponding to the well-known tradeoff between the thickness and velocity of the LVZ. Two common-sense constraints can be applied, however, to greatly reduce the range of possibilities. Points having τ's larger than a certain value can be rejected, since they will fail to correct the negative thickness, Z_n originally produced in the inversion. For any given value for p_{LVZ}, the minimum thickness for the overlying layer is that which gives $\Delta\tau_{n+1}(p_{n+1}) = 0$. This value is a function of the previously determined $V(Z)$ model down to the $n-1$th layer, and by τ_{LVZ}, the latter value can be found by algebraic manipulation of Equations (9) and (10). The points satisfying this criterion form a maximum bound for the synthetic points (Figure 15). For any chosen p_{LVZ}, a τ_{LVZ} lying on this bound provides the maximum allowable thickness of the HVC and depth to the bottom of the LVZ. All points with a greater τ_{LVZ} value will fail to correct the negative thickness that flagged the existence of the LVZ in the first place.

A similar analysis can be carried out to find the τ_{LVZ} that will produce the minimum allowable HVZ thickness and LVZ depth for any particular p_{LVZ}. Points along the resulting minimum bound produce an HVC thickness of zero, as whon in the examples of Figure (16). Any point with a smaller τ_{LVZ} value will produce an HVZ with negative thickness.

Any point lying between the upper and lower bounds thus defined will satisfy the two limiting constraints (Figure 17). Depending on the data, the corresponding $V(Z)$ bounds may be disturbingly large. The bounds may be tightened somewhat by imposing the constraint that the HVC have some finite thickness,

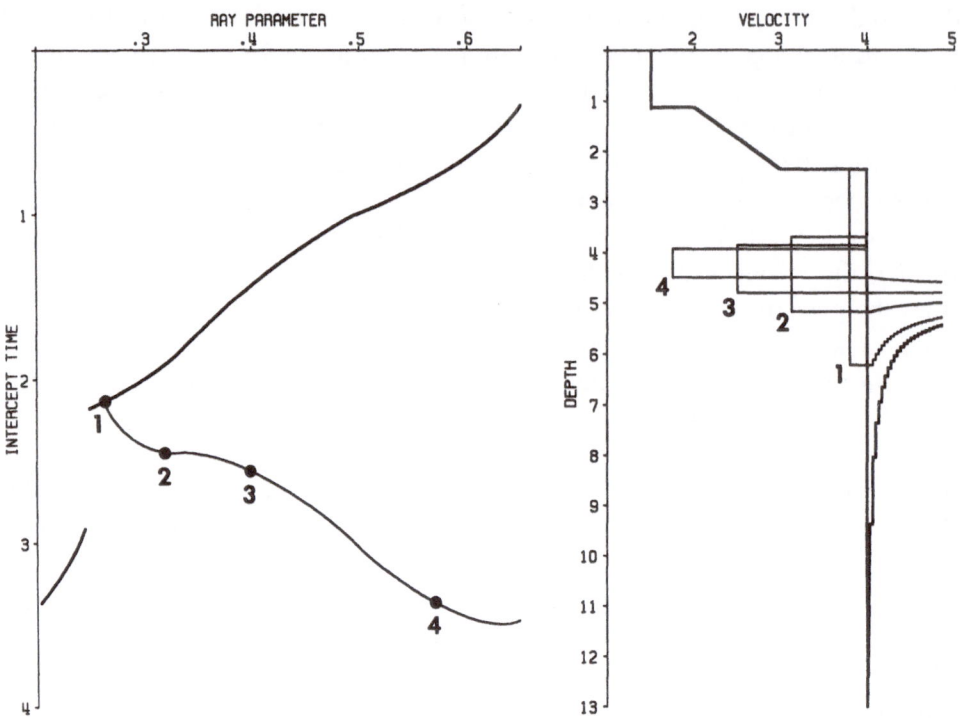

Fig. 15. A simple model with a LVZ layer is raytraced to produce a well sampled (dp = 0.002 sec/km) set of $\tau(p)$ points. The LVZ gap produces a characteristic response in the Tau-sum inversion (compare with Figure 2). The maximum bound for synthetic $\tau(p)$ points that correct the first negative thickness for a given LVZ slowness is shown on the left, and the inversion results for the four marked points are superimposed on the uncorrected result. The correct velocity function is returned by the inversion with point 3. The LVZ with the highest possible velocity (1) produces a zero thickness above the LVZ, as well as below it. In this case, the choice of this velocity corresponds exactly to correction by removing $\tau(p)$ picks (Figure 4).

or loosened by allowing small negative thickness layers above and/or below the LVZ, corresponding to the expected 'slop' due to possible sampling jitter. The point at which the two bounds intersect defines the LVZ which has the maximum possible velocity and which violates neither of the two criteria. This point can always be calculated automatically. When the HVC is an homogeneous layer, as in Figure 17, the result is identical to the removal of (supposedly) precritical points. In cases where there are well-sampled velocity gradients immediately above the LVZ, the automatic correction can be useful in producing a 'hands off' reasonable value for the LVZ, and for the depth to the underlying layers. In Figure 18 two LVZ's can be seen in initial Tau-sum inversions. They are removed by application of the automatic correction. This 'quick-and-dirty' LVZ correction compares well with results from forward modeling and synthetic seismogram modeling.

Even if it is known that an LVZ is responsible for a postcritical data gap,

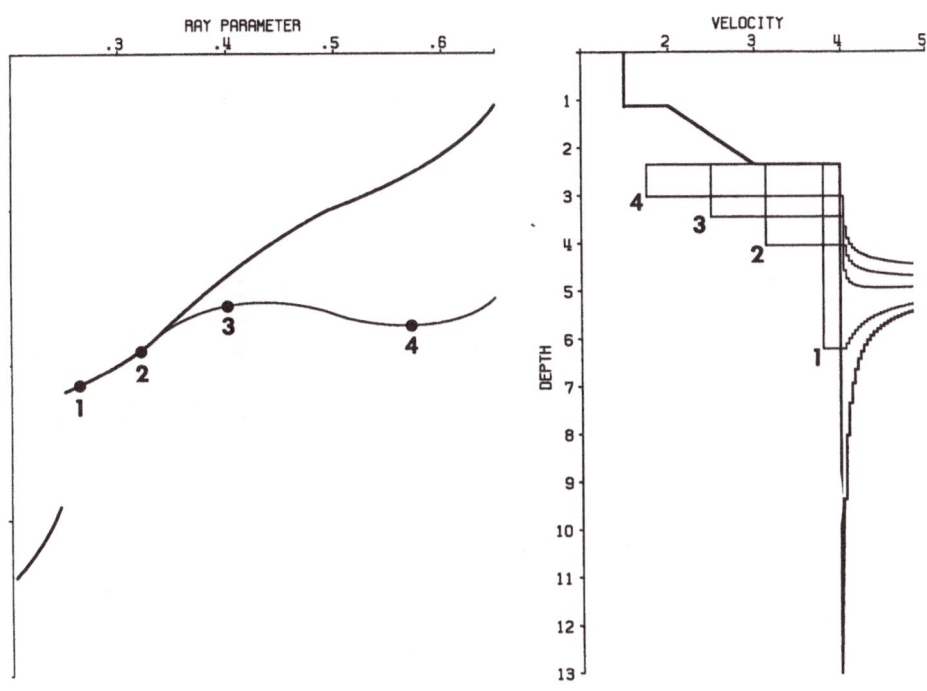

Fig. 16.   The minimum bound for the range of allowable LVZ $\tau(p)$ points (left), produces Tau-sum inversions with zero thickness for the overlying layer (right).

obtaining the correct $\tau(p)$ values to insert into the Tau-sum inversion may not always be this simple. When precritical reflections can be detected from above and below the LVZ, $\Delta\tau^2 - p^2$ analysis will give a velocity and thickness for the zone. In the case that reflections are seen only from the top of the LVZ (bottom of the HVC), the HVC thickness and velocity can be calculated, assuming that it is a homogeneous layer, by a hybrid method, in which $\Delta\tau^2 - p^2$ analysis is applied after stripping the overlying layers determined by Tau-sum recursion. Knowledge of the HVC thickness can greatly reduce the range of possible LVZ's.

   If reflections are seen only from the bottom of the LVZ, then it is possible to uniquely determine the best-fitting homogeneous HVC–LVZ pair of layers. This can be done via an extension of the 'τ NMO' described by Stoffa *et al.* (1981). Tau NMO is simply the application of the p dependent time shift $\tau_n(0)-\tau_n(p)$. As we have seen, the inclusion of any possible τ_{LVZ} point results in the construction of a particular pair of HVC and LVZ layers which can be used to predict the τ NMO shifts for the reflection from the bottom of the LVZ. The point producing the maximum stack and semblance at the predicted $\tau_{LVZ}(0)$ normal time will also yield the HVC–LVZ sequence best fitting the reflected arrivals. This technique is illustrated in Figure 19. Postcritically reflected and refracted points (left) have been picked down to an obvious LVZ. Picks from a well-defined

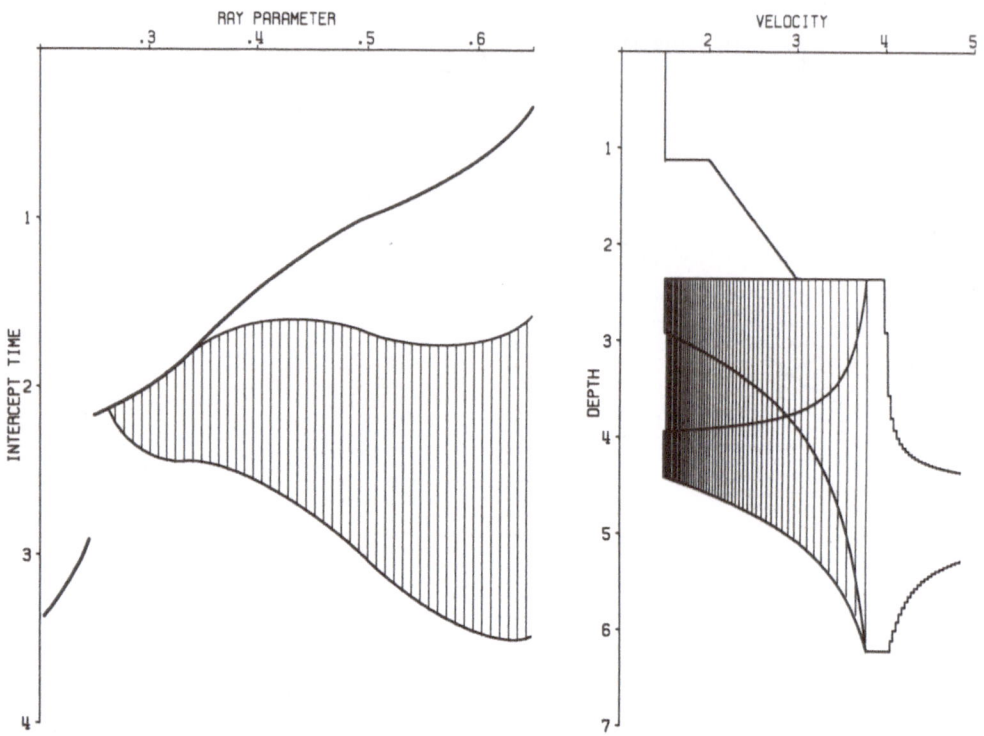

Fig. 17.   The entire field of allowable synthetic LVZ $\tau(p)$ points (left) and the corresponding Tau-sum results. The $V(Z)$ envelopes corresponding to the minimum and maximum LVZ $\tau(p)$ bounds are shown on the right. The entire field of possible points is defined by thin vertical lines, equally spaced in p. For each of these there is a corresponding line on the right. Following the minimum LVZ $\tau(p)$ bound minimizes the LVZ depth, as well as the depths to layers below. The inverse effect is obtained by following the maximum bound.

reflection from the bottom of the LVZ have also been determined. The 'raw' Tau-sum recursion results (right) show the characteristic LVZ response. Minimum and maximum τ_{LVZ} bounds have been plotted, and the 'minimum correction' Tau-sum results calculated. The residual curvature in the reflection horizon depths indicates that the simple automatic LVZ correction is insufficient to completely remove the LVZ response. When the points just beyond the LVZ-induced $\tau(p)$ data gap correspond to arrivals reflected from a single interface, reason dictates that the correct synthetic $\tau_{LVZ}(p_{LVZ})$ point must lie along the maximum τ_{LVZ} bound; the locus of all points that corrects the first negative thickness in the Tau-sum recursion. A trial error τ NMO and stack has been performed for each possible point along the maximum τ_{LVZ} bound in Figure 19. The maximum value of coherence (semblance x stack) for each point is plotted below the $\tau(p)$ picks. There is an obvious peak in the coherence curve near $p_{LVZ} = 0.4$ sec/km. The corresponding Tau-sum recursion result, shown on the right, exhibits the desired (nearly) flat behavior for the sub-LVZ reflector.

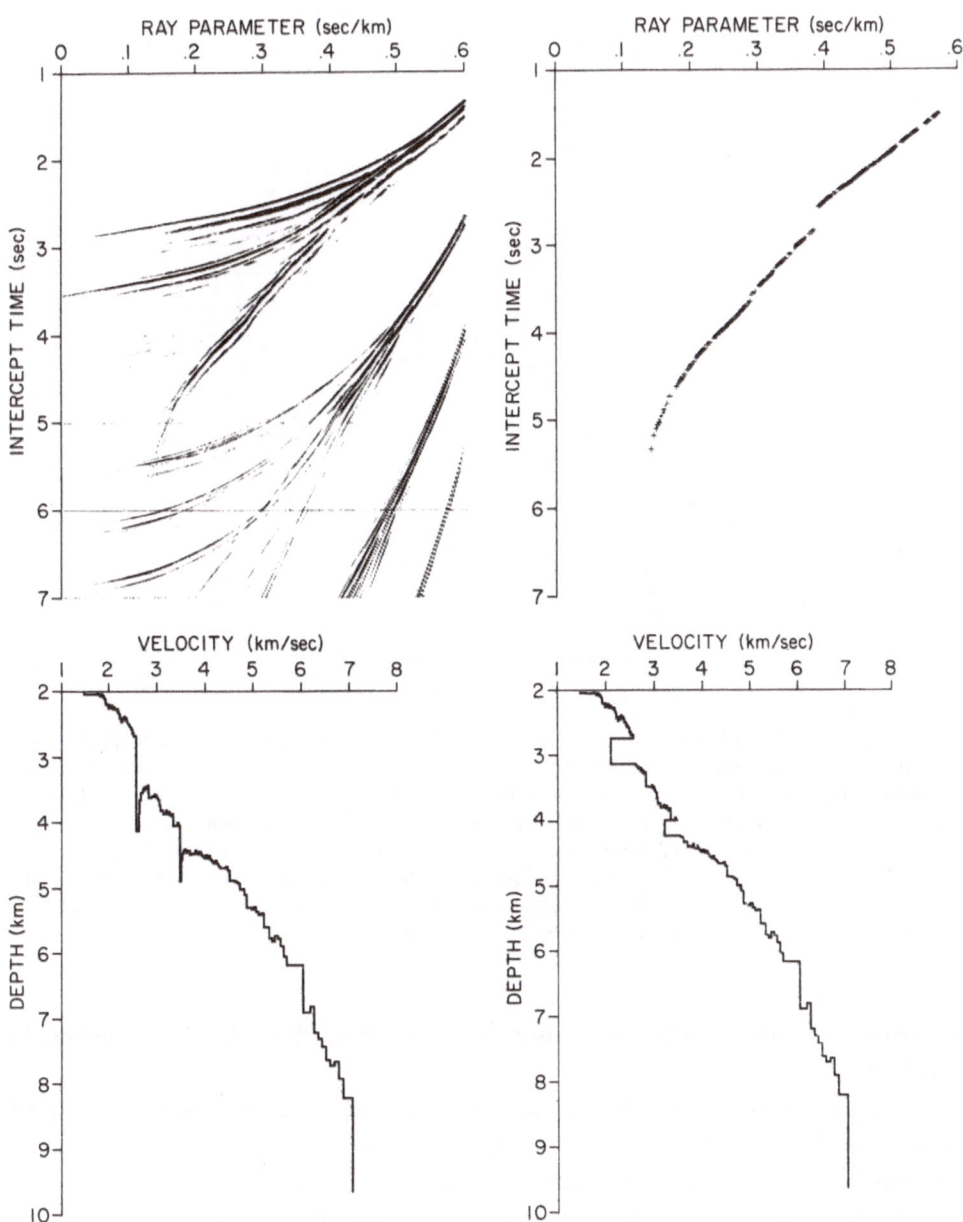

Fig. 18.   An example of automatic LVZ correction in a case where two LVZ's affect the initial Tau-sum inversion. The effects of finite time sampling are also apparent.

LIMITATIONS ON PRECRITICAL INVERSIONS

It is obvious that $\Delta\tau^2 - p^2$ inversion can be quite useful, especially when there are missing postcritical data, or LVZ's. It is important, however, to understand the limitations of this approach (even before examining the effects of dip, which is

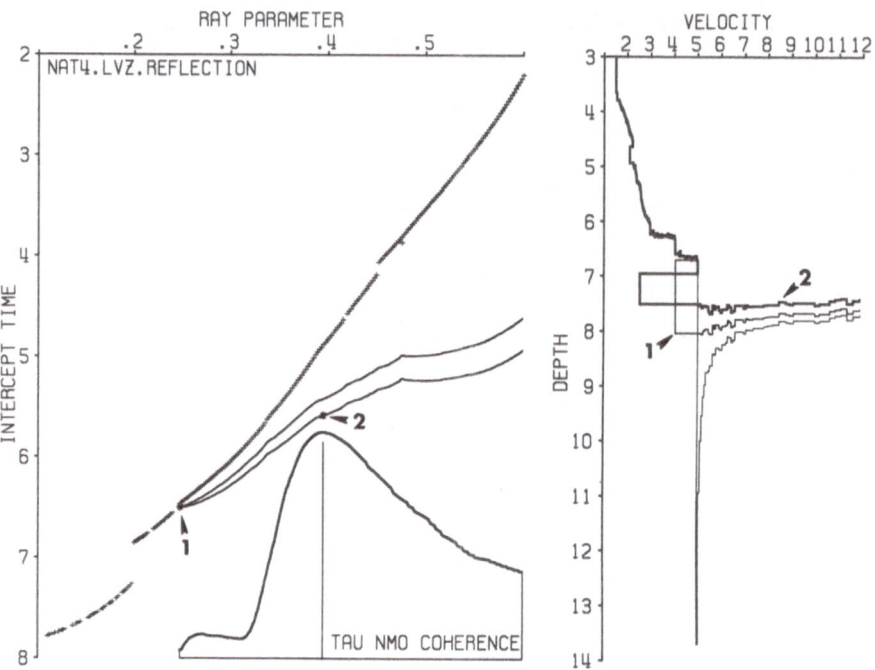

Fig. 19.   Postcritical $\tau(p)$ points from above a large LVZ, and precritically reflected points from below the LVZ (left) produce a characteristic Tau-sum response (right). τ_{LVZ} bounds were calculated and plotted with the $\tau(p)$ picks. The automatic LVZ correction (1) fails to flatten the Tau-sum solution for the bottom of the LVZ. τ NMO was calculated for every point along the maximum τ_{LVZ} bound, and coherency measured for the stacked data when moved out accordingly. The maximum coherence is plotted as a function of p_{LVZ}, just below the τ_{LVZ} bounds. This curve has a clearly defined peak near $p_{LVZ} = 0.4$. The corresponding Tau-sum solution (2) shows the best possible flattening of the sub-LVZ horizon, indicating that the LVZ velocity is about 2.5 km/sec.

inappropriate at this point). The limiting effects can be illustrated by some simple modeling.

The resolution of precritical $\tau(p)$ inversion is constrained by the bandwidth of the data, and by limitations on the range of ray parameters over which a particular event can be tracked. The first effect depends on the proportion of the $\Delta\tau$ resolution, $\delta\tau$, the layer's slowness, and its interval two-way time, as shown in Figure 20, which shows how the uncertainty maps into $\Delta\tau^2 - p^2$ space. The $\delta\tau$ uncertainty envelope is quite small for ray parameters near the layer slowness, u. When these ray parameters are achieved in the data, the velocity resolution can be quite good, while the resolution in interval two-way time $\Delta\tau(0)$ is unchanged.

The bandwidth-induced resolution loss is compounded, however, when the data are limited to small values of p, where the change, $d\Delta\tau/dp = -p(2Z)_2/\Delta\tau$, varies quite slowly. This effect can be seen in Figure 11(c). In the case of an LVZ, of course, $\Delta\tau$ never goes to zero, and 100% ray parameter coverage is never

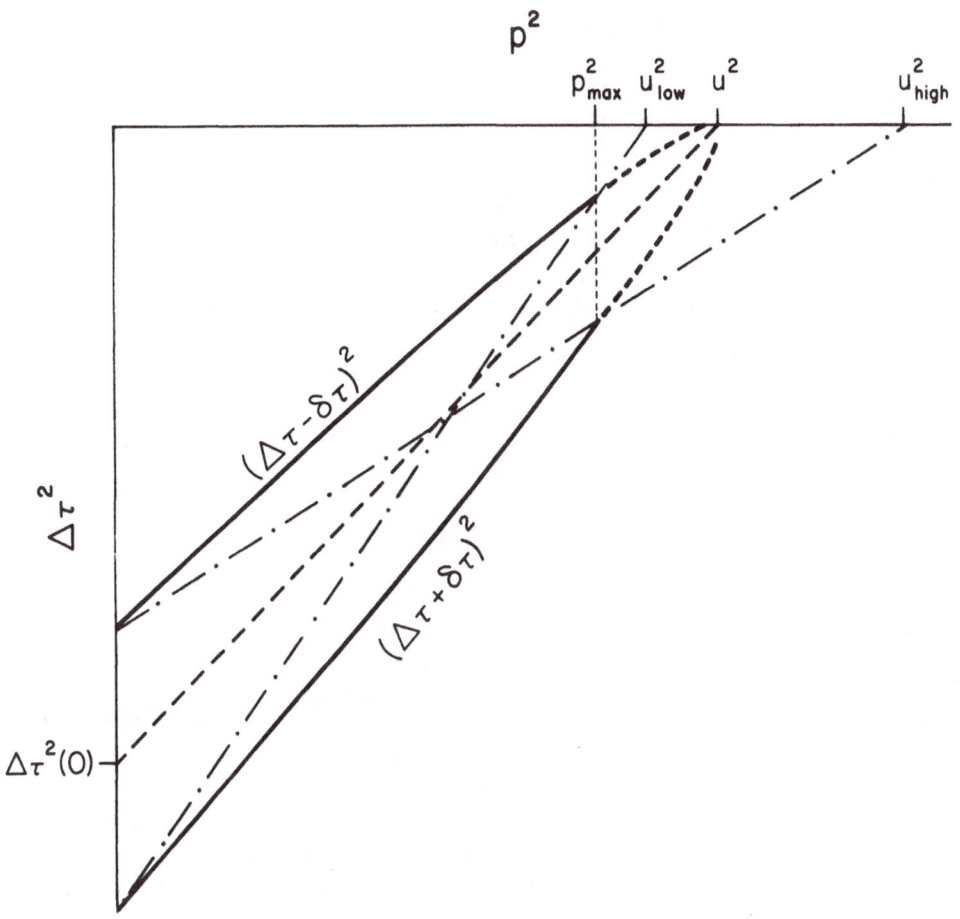

Fig. 20. A time uncertainty, $\delta\tau$, which is constant at all p's forms an envelope in $\Delta\tau^2$–p^2 space. The envelope is shown for a layer with slowness u and two-way interval time $\Delta\tau(0)$. Assuming that all ray parameters have been sampled between 0 and p_{max}, the minimum (u_{low}) and maximum (u_{high}) bounds on slowness resolution can be easily determined.

attained. The homogeneous layer assumed by the $\Delta\tau^2 - p^2$ method implies that zero $\Delta\tau$ will only be seen at infinite offset. Figure 21 shows, for a layer with typical slowness ($u = 0.5$ sec/km) and an excellent $\delta\tau$ (4 msec), the increase in resolution as higher p's are sampled, for a number of layer thicknesses. It can be seen, in Figure 20, that the line through the $(\Delta\tau \pm \delta\tau)^2$ bounds defining maximum layer slowness will become horizontal when a particular ray parameter cutoff is reached. The curves in Figure 21 are terminated at the corresponding p's.

The fact that the curves in Figure 21 will converge to zero error at p_{max} for a homogeneous layer anywhere within a more complex structure, shows the inherent superiority of $\Delta\tau^2 - p^2$ methods over the traditional $T^2 - X^2$ techniques, as described by Shultz (1982) and Stoffa *et al.* (1982). Unfortunately, other errors in $\Delta\tau^2 - p^2$ analysis may arise from incompatibilities that may exist between the model assumptions and the real earth.

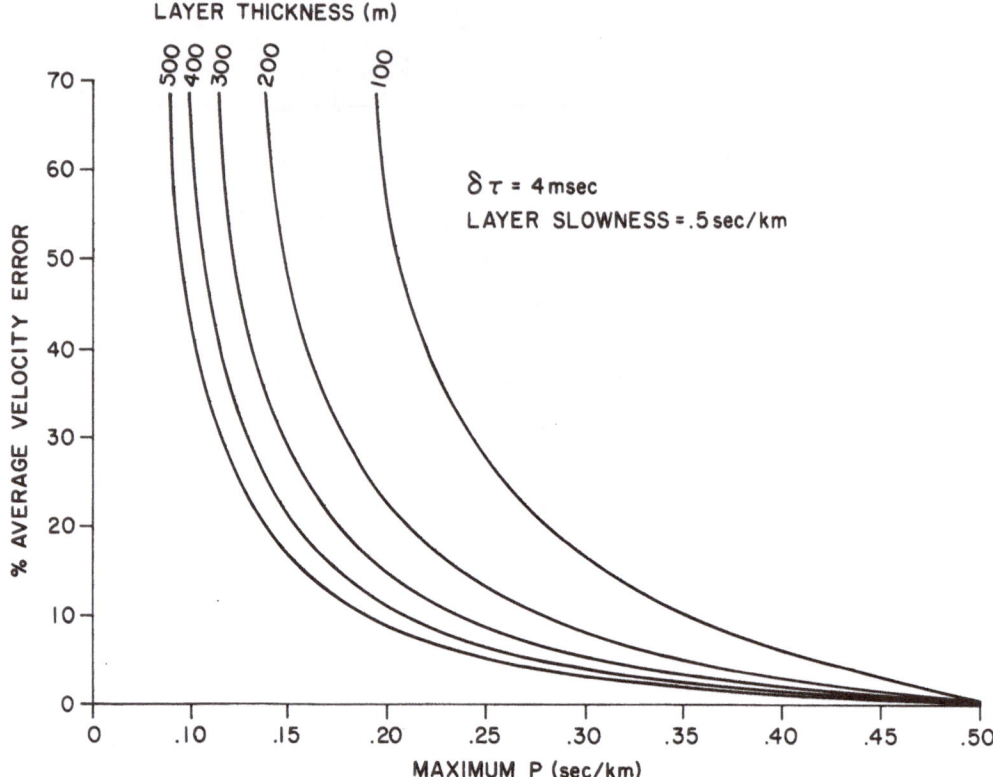

Fig. 21.   For a fixed $\delta\tau$ = 4 msec, and a layer slowness u = 0.5 sec/km, the effect of a limited range of observable p's on slowness resolution in $\Delta\tau^2-p^2$ inversion is shown for a number of layer thicknesses varying between 100 and 500 m.

When a single, homogeneous layer is involved, postcritical arrivals can be included in the $\Delta\tau^2 - p^2$ inversion without penalty (neglecting problems due to phase changes). When the layer's velocity function is a gradient, however, this is no longer true (Hake *et al.*, 1986). Figure 22 shows the effects of $\Delta\tau^2 - p^2$ inversion when the postcritically refracted arrivals from a gradient layer are included (a situation that would be avoided, under Schultz's (1982) method, due to low semblance values along this path), and when only the precritical arrivals are used. In the latter case, the correct thickness and the average velocity are obtained. To detect the velocity gradient, however, it would be necessary to compare the results with reflectivity modeling and/or to closely examine the apparent changes in thickness and velocity as larger offset data were included.

When, for some reason, a stack of homogeneous layers is represented by reflections only from the top and bottom of the stack, an offset-dependent error is induced into $\Delta\tau^2 - p^2$ inversion which is identically equivalent to the offset bias in the use of Dix's equation in the analysis of stacking velocities (Al-Chalabi, 1974). Figure 23 shows this effect for a single set of offsets. A simple, but

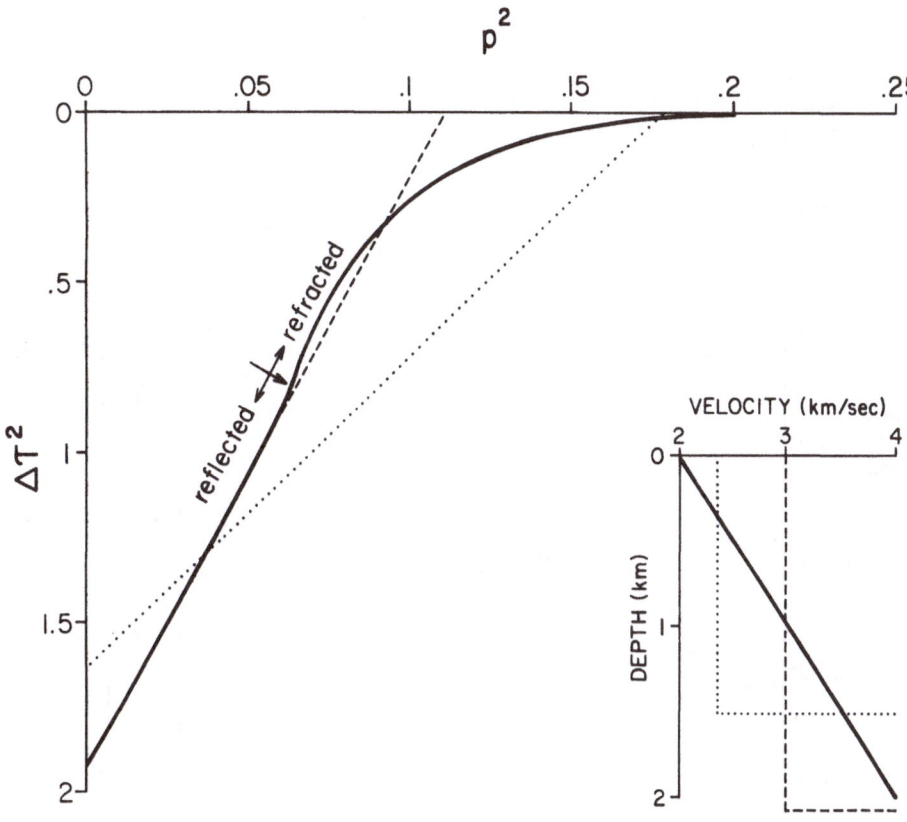

Fig. 22. When $\Delta\tau^2-p^2$ inversion is applied to reflected and refracted arrivals from a gradient layer (solid line, lower right), the results range from reasonable, when only precritical reflections are used (dashed lines) to poor, in the unlikely event that postcritical arrivals are included, as well (dotted lines). In either case, the gradient is not resolved.

vertically heterogeneous two-layer model produces a curved $\Delta\tau^2 - p^2$ path, the analysis of which will produce single-layer values of velocity and thickness which will steadily increase with increasing offset. Again, close examination of the changes in these values with offset might allow the determination of the degree of heterogeneity, using either linear algebraic inversion of a system of equations like (10), or *ad-hoc* methods (Al-Chalabi, 1974). Figure 24 shows, however, that precritical traveltime analysis will be unable to discriminate the difference between: a series of layers with alternating velocities, a simple high-low velocity sequence, or a simple low-high sequence.

Two-Dimensional Inversion

When the plane interfaces separating the homogeneous layers of the one-dimensional model described in the previous section are allowed to dip along a common strike direction, it is convenient to designate the slownesses of up- and

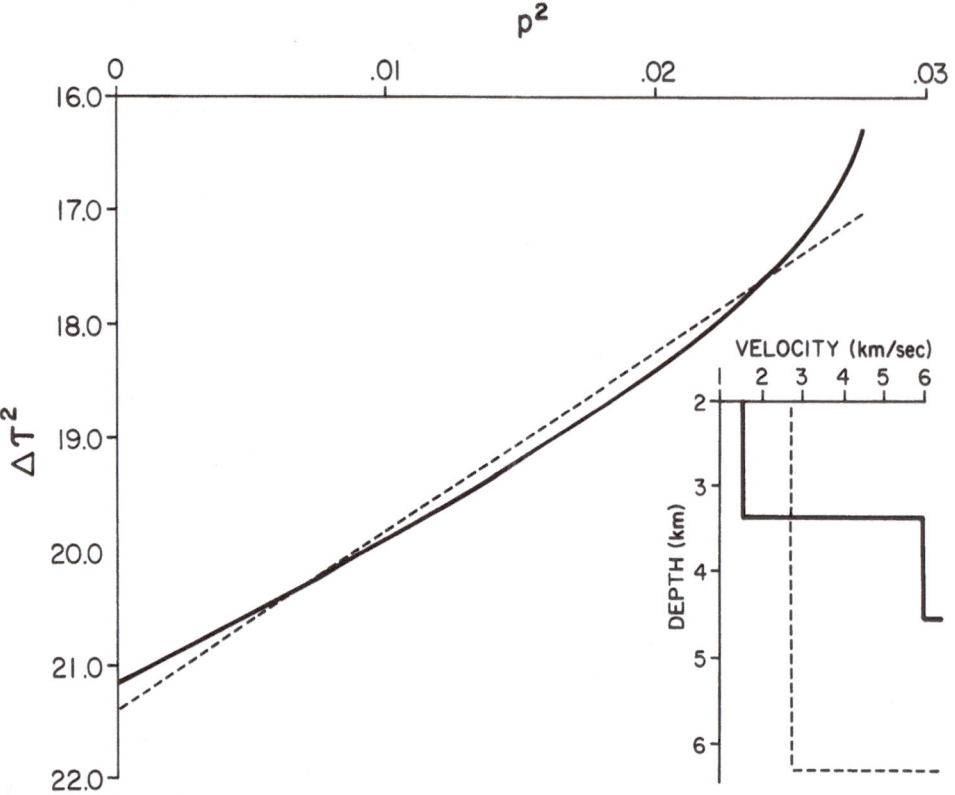

Fig. 23. When $\Delta\tau(p)$ is determined for an interval in which a bimodal velocity distribution is present (solid lines), the $\Delta\tau^2-p^2$ curve is not straight, and the inverted layer parameters will depend on the p coverage.

downgoing rays by suffixes $(p_a, q_{aj}, p_b, q_{bj})$, and the two-dimensional version of the generalized traveltime equation (7) can be written

$$T = X_a p_a + X_b p_b + \sum Z_j q_{aj} + \sum Z_k q_{bk}. \tag{12}$$

In this case, the intercept time sums of the up- and downgoing rays are separated to include the case in which source and receiver are in different layers.

Inversion methods for postcritical rays typically revolve around the determination of p_a and p_b, the apparent horizontal slownesses at either end of a given ray. Knowing dips and slownesses in overlying layers, Snell's law is used to find the true angles of incidence for the ray at the critical interface and, therefore, the true refraction/reflection velocity and layer dip. These values are retained for future use, and the ray's vertical slownesses are combined with the two intercept times to determine the thickness above the refractor/reflector.

Since simultaneous measurement of the apparent horizontal slownesses at both ends of a ray is very rare, it is usually necessary to identify 'congruent' rays; those that have identical incidence angles, but traverse different parts of the section. This process has been described for surface sources and receivers in the interpre-

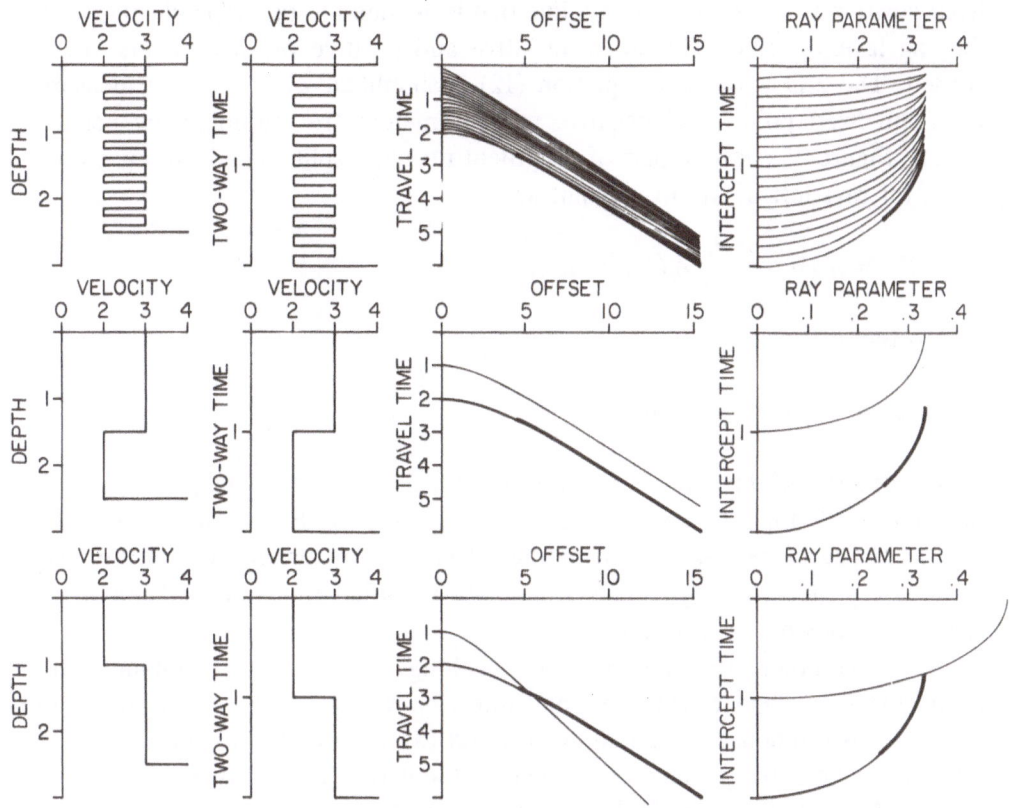

Fig. 24.   Three different models whose basal reflections (heavy lines) have the same arrival times and intercept times.

tation of $T(X)$ data from reversed spreads (Ewing *et al.*, 1939) and split spreads (Johnson, 1976). As in the application of the classical one-dimensional slope-intercept techniques, the traveltime slopes and intercepts have traditionally been interpreted as corresponding to headwaves for purposes of inversion with these methods. Diebold and Stoffa's (1981) generalization of the traveltime equation, and the Tau-sum inversion, to postcritical arrivals of any kind still holds, however, and the traditional techniques are easily adapted to densely sampled $\tau(p)$ data sets.

Milkereit *et al.* (1985) showed how inversion for reversed and split spread profiles could be accomplished in the $\tau(p)$ domain, using Equation (12), and its implied equivalence of headwave, refracted, and postcritically reflected arrivals. We present a modification of their technique, which greatly facilitates the identification of congruent rays for reversed profiles.

SPLIT SPREAD PROFILES

During the acquisition of a split spread profile, the receiver (or source) remains in a fixed position, while the sources (or receivers) are moved from one side of the

fixed position to the other, along a line that is assumed to be perpendicular to the
dipping layers' strike. Therefore, negative and positive ray parameters are recorded. The 2-D traveltime equation (12) is simplified when offset is measured
from the fixed point, and negative ray parameters are multiplied by negative
offsets. Figure 25 shows a pair of congruent rays in a split spread survey. Traveltimes for rays left of the fixed point are

$$T_A = X_A p_a + \sum_{j=1}^{4} Z_j(q_{aj} + q_{bj}),$$

and to the left;

$$T_B = X_B p_b + \sum_{j=1}^{4} Z_j(q_{aj} + q_{bj}).$$

In the presence of dip, the observed horizontal slownesses, p_a and p_b can be quite
different, while the intercept times for congruent rays in the two directions will be
identical. Therefore, correlation of the congruent ray slownesses can be easily
made for postcritical $\tau(p)$ picks from data recorded on either side of a split
spread, as shown in Figure 26.

Once a complete set of $\tau(p_a)$ and matching $\tau(p_b)$ picks are obtained, true
layer slownesses and thicknesses, and interface dips are obtained from the top
down. In two dimensions, Equations (2) and (3) are greatly simplified, since the
layer normal can be specified by its angle to the vertical, the dip; and a ray can be
described uniquely by a single horizontal slowness. Inversion consists of two
steps; first, true slownesses and dips are obtained by propagating the observed

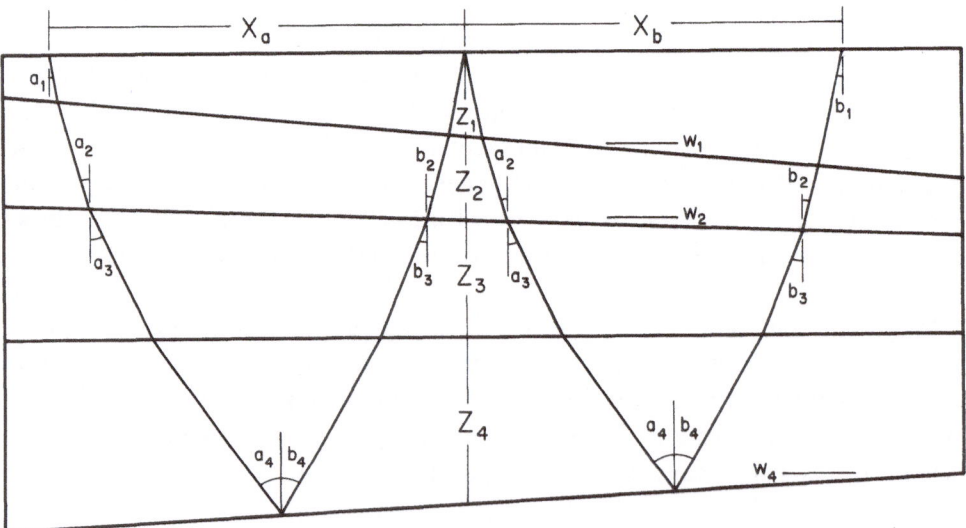

Fig. 25.   Congruent rays in a split spread profile. Each layer has dip, $w_j(w_3 = 0.)$. The offsets X_a and
X_b are different, though each ray has the same set of angles and slownesses.

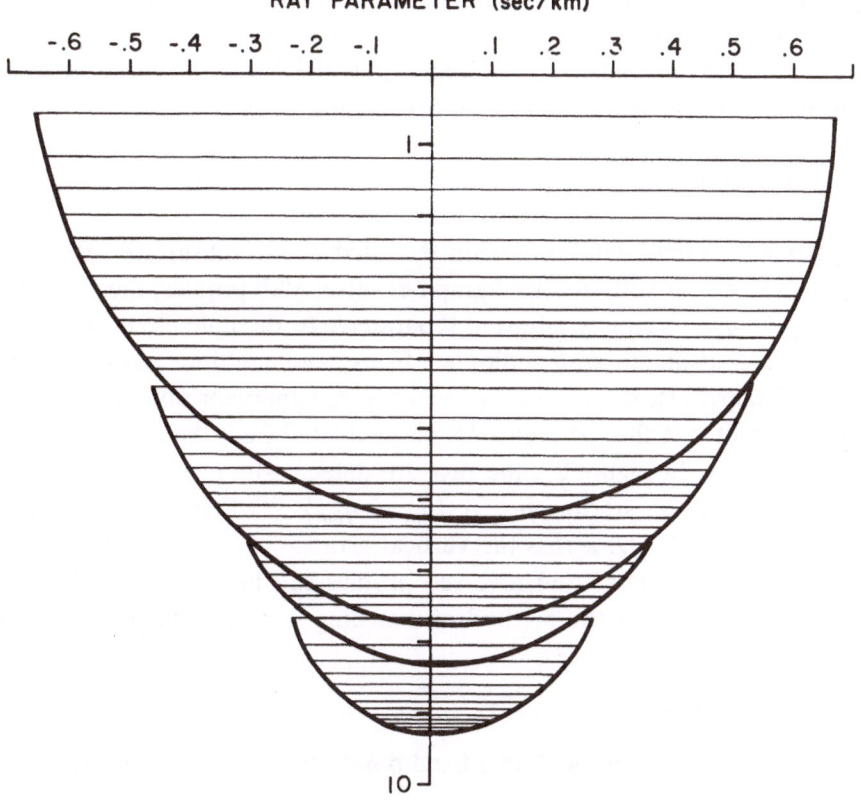

RAY PARAMETER (sec/km)

Fig. 26. $\tau(p)$ plots for the split-spread profile geometry of Figure 25. Negative ray parameters correspond to negative offsets. The horizontal lines connect postcritical $\tau(p)$ points for congruent rays.

slownesses downward through the appropriate number of interfaces. Starting with the angles

$$a_1 = \sin^{-1}(u_1 p_a) \quad \text{and} \quad b_1 = \sin^{-1}(u_1 p_b),$$

the algorithm

$$a_{n+1} = \sin^{-1}[u_{n+1} \sin(a_n + w_n)/u_n] - w_n$$

is used to correct the effects of known overlying structure. Here w_n is the dip of the interface separating the layers with slownesses u_n and u_{n+1} (Figure 25). The vertical slownesses $q_{aj} = u_j \cos a_j$ are accumulated during the downward propagation, for use later. When the signs of the rays are correctly applied, the same algorithm can be used for the other set of rays, with b's substituted for the a's. When the interface (with index, say, k) of postcritical refraction or reflection is reached, the critical angle i_c is given by $i_c = (a_k - b_k)/2$, and the dip by $w_k = (a_k + b_k)/2$. It may appear that these signs are reversed, but remember that a_k and b_k have opposite sign. The true apparent velocity below the interface is u_{k+1}

$= u_k \sin(i_c)$. If these formulas seem familiar, no wonder; they appear in innumerable texts on the refraction method.

The second inversion step is to determine the layer thickness by the two-dimensional equivalent of the Tau-sum recursion, with

$$\tau_n = \sum_{j=1}^{n-1} Z_j(q_{aj} + q_{bj}). \tag{13}$$

When the split spread data have been recorded with a subsurface receiver, as in the case of an ocean bottom seismograph or offset VSP profiles, the starting index for the q_{aj} or q_{bj}'s is simply adjusted separately. In the unlikely event that offset VSP data are available with multiple offsets, and multiple downhole receiver locations (Figure 27), the problem of identifying congruent rays is trivial, though the ray parameter at the receiver must be calculated from the measured apparent vertical slowness. In Figure 27, the rays are detected by receivers in a layer with slowness u_m, and the starting $p_{bm} = (u_m^2 - q_{bm}^2)^{1/2}$, where q_{bm} is the apparent vertical slowness, dT/dZ across the vertical array.

Inversion of $\tau(p)$ data from reversed profiles can be carried out in the identical manner, except for the method of matching p_a and p_b for congruent rays.

REVERSED PROFILES

In a reversed profile, data with two fixed points are recorded; the fixed point for

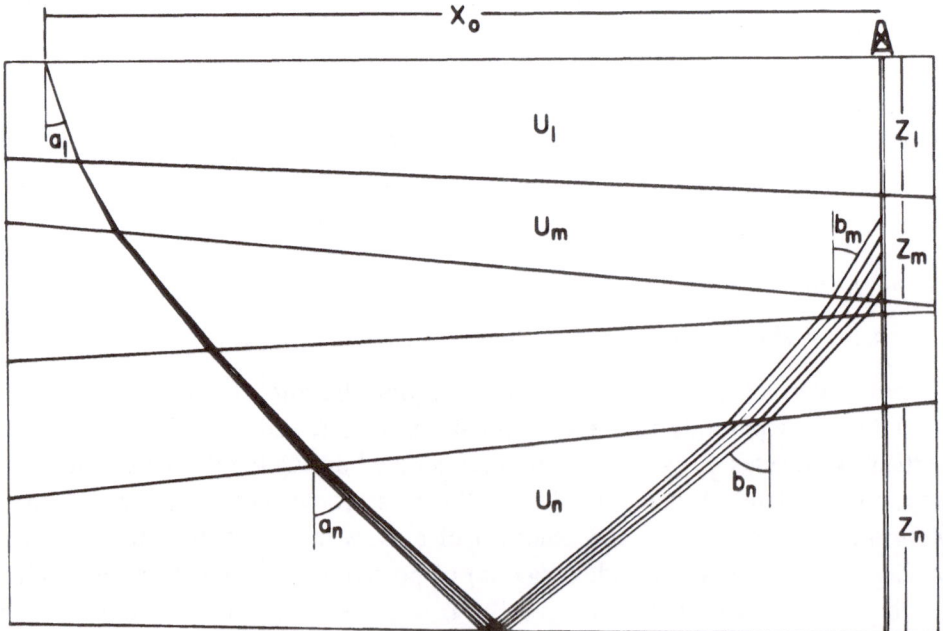

Fig. 27.   A multi-offset, multi-receiver VSP profile in two dimensional structure.

rays to the left is located towards the right, and the other is at the left, for the rays to the right. Figure 28 shows a set of congruent rays for reversed profile geometry. If offsets, X_a and X_b, for the two sets of rays are measured from the appropriate fixed points, both the apparent horizontal slownesses and the intercept times for congruent rays will differ

$$T_a = X_a p_b + \sum Z_{aj}(q_{aj} + q_{bj}) = X_a p_b + \tau_a,$$

and

$$T_b = X_b p_a + \sum Z_{bj}(q_{aj} + q_{bj}) = X_b p_a + \tau_b.$$

Although these intercept times are likely to be different for the two congruent rays in the reversed profiles, two other values will be equal; what are sometimes called the 'reverse times' (Ewing, 1963);

$$T_{aR} = T_a + p_b(X_{off} - X_a) = \tau_a + p_b X_{off},$$

$$T_{bR} = T_b + p_a(X_{off} - X_b) = \tau_b + p_a X_{off}.$$

That these two values must be equal is easily proved in the single layer case

$$T_{aR} - T_{bR} = X_{off}(p_b - p_a) + (Z_a - Z_b)(q_a + q_b).$$

Since $Z_a - Z_b = X_{off} \tan w$, and Equation (3) reduces to

$$p_a - p_b = (q_a + q_a) \tan w,$$

this is obviously true.

The time shift required to determine the 'reverse time' is linear in the $\tau(p)$

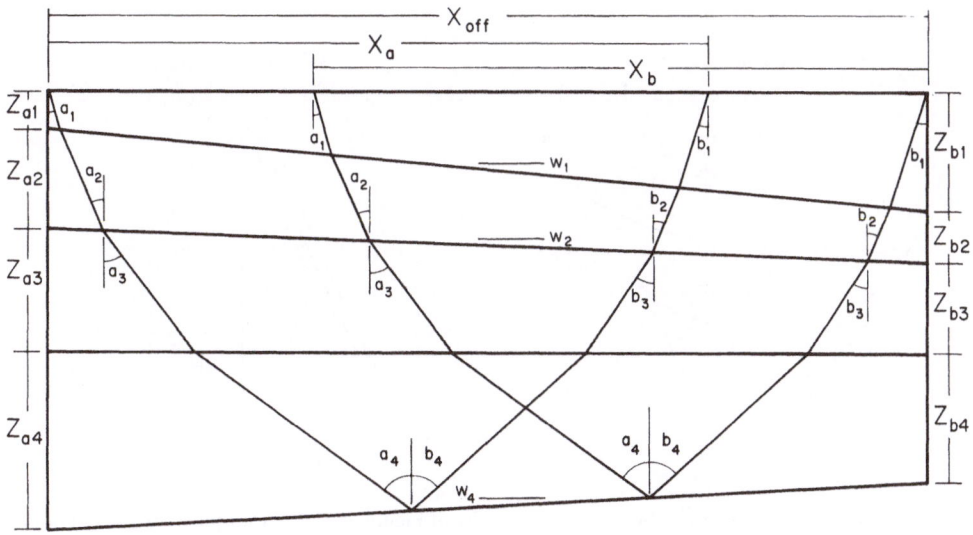

Fig. 28. Congruent rays in a reversed spread profile.

domain, with the apparent horizontal slowness multiplied by the station offset, X_{off}. Therefore, if the $\tau(p)$ transforms of the two reversed profiles have the correct linear moveout applied, congruent rays will have identical 'intercept' times, as shown in Figure 29. Since the required shift is linear, the original τ's can be recovered, and Tau-sum recursion of Equation (13) can again be applied. The correlation of p_a and p_b for congruent rays can be made, as described by Milkereit *et al.* (1985), without this linear shift, but then instantaneous slopes along the trajectories of the picked $\tau(p)$ points must be calculated. This process is somewhat more sensitive to errors in picking.

TWO-DIMENSIONAL PRECRITICAL $\tau(p)$ INVERSION

The equality of intercept times, by themselves, in the case of a split spread profile, or reduced appropriately in a reversed profile, applies just as well to precritically reflected arrivals as it does to the postcritical arrivals treated above. As with Shultz's (1976, 1982) various treatments of the inversion of reflections in one-dimensional $\tau(p)$ data, two basic approaches are possible. The first is the minimal/manual method, where the minimum necessary number of $\tau(\pm p)$ picks, manual or otherwise, are used to define the three parameters (slowness, thickness, dip) defining a single layer. The second involves a 3-D computed semblance scan, analogous to that used in Shultz's (1982) 'Unravel'.

Using Equation (3), it can be easily shown that for a reflection in a single, two-dimensionally dipping layer,

$$\Delta \tau = 2Z \cos(w) \cos(i_c)/V;$$

i_c being, as before, the reflection angle, w the dip. It is also easily shown that the

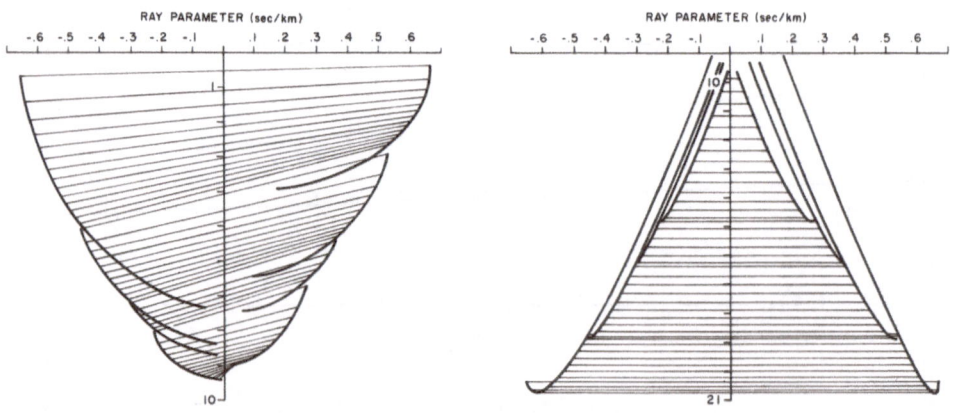

Fig. 29. The $\tau(p)$ plot for the reversed spread profile of Figure 28 (left), with postcritical $\tau(p)$ points for congruent rays connected. In reversed profiles, these τ's are not equal. However, when the linear shift $\tau(p) = \tau(p) + pX_{\text{off}}$ is applied (right), it is a trivial matter to correlate congruent arrivals.

sum of the negative and positive ray parameters for the congruent ray is

$$p_a + p_b = 2 \sin(w) \cos(i_c).$$

Therefore, for congruent rays with the same (reduced, if necessary) intercept time,

$$\Delta\tau = Z(p_a + p_b)/V \tan(w).$$

When three such $(p_a + p_b)$ observations are made, these will be sufficient to solve for the three values, V, Z, and $\tan(w)$.

In a single layer underlain by a dipping interface, the intercept time can be related to the apparent horizontal slowness $\pm p$ by the reflection angle; $\pm p = \sin(i_c \pm w)/V$. In Schultz's (1982) implementation of the 1-D 'Unravel' method, a semblance search is made in terms of $\Delta\tau(0)$ and V, for the ellipse corresponding to the reflection from a single layer. A similar search can be made in 2-D, except that three variables must be scanned. With the equations above, they are Z, V, and w. For each V, w pair to be considered, $i_c = \sin^{-1}(pV) \pm w$ is used to shift each intercept time $\tau(p)$ trace to the corresponding $Z = \tau V/2 \cos(w) \cos(i_c)$. A vertical stack and semblance calculation across $\pm p$ for all of the resulting $Z(p, V, w)$ traces will produce a three-dimensional $Z(V, w)$ semblance map with a 'peak' at the appropriate spot. Obviously, $T(V, w)$, $T(u, w)$, or $Z(u, w)$ can be calculated instead, if desired.

Once values for the current layer have been chosen, they can be used to strip the subsequent data of the layer's τ contribution, and the search recommenced. In the 2-D case, however, there is a ray parameter shift across the interface;

$$p_{j+1} - q_{j+1} \tan(w_j) = p_j - q_j \tan(w_j) \quad \text{(Diebold, 1987)}.$$

This relationship can also be extracted from Equation (2). Since q_{j+1} has a quadratic dependence on p_{j+1}, the solution for the new apparent slowness has a somewhat complicated form. Since q_{j+1} also depends on the (as yet undetermined) slowness u_{j+1}, it must be calculated and applied for each V element in the next semblance search. Once the next set of Z, V, w parameters is chosen, the correction can be made once and for all, so that the complexity of the operation does not increase for deeper layers.

Three-Dimensional $\tau(p)$ Data

Most of the one-dimensional $\tau(p)$ transformation methods now in use can be adapted for use in obtaining quasi three-dimensional $\tau(p)$ data by repeated application along horizontal lines within three-dimensional $T(X, Y)$ data sets. Since such data sets are commonly acquired with geometries consisting of multiple 2-D lines of data coverage, this is a sensible approach. One transform method is to make repeated 2-D stacks of data collected along radial tracks $T = \langle X, Y, 0 \rangle$, to obtain $\tau(p_R)$ [where $p_R = (p_x X + p_y Y)$]. When true 3-D $T(X, Y)$ data are available, $\tau(p_x, p_y)$ can be obtained by some variation of the 2-D Fourier transform techniques currently in use (Phinney *et al.*, 1981; Treitel *et al.*, 1982, among others), but with a triple transform.

No matter what profile geometry is used, the accumulated vertical delay terms will have virtually identical form, Equation (8). In the typical case that sources and receivers are located on a common horizontal surface, which consists of the upper limit of layer #1, the intercept time is given by Equation (13). Although this 3-D intercept time equation is identical in form to that used in 2-D inversion, the q_{aj}'s and q_{bj}'s each depend on two horizontal slownesses.

Three-Dimensional τ Inversion

When a three-dimensionally dipping model is to be used in homogeneous layer inversion, each layer is still characterized by a single slowness and thickness, but two additional variables are required to describe the dip and strike of the underlying interface. We have seen that a single postcritical $\tau(p)$ observation is sufficient to determine velocity and thickness for a single one-dimensional layer, and that two observations are required for a single two-dimensional layer. It follows that three postcritical $\tau(p)$ observations will be required to define the thickness, velocity and attitude of a layer in three dimensions.

As in the 2-D case, the crucial step in 3-D inversion is the correlation of the $\tau(p_x\, p_y)$ values for each set of three congruent rays. In a seismic profile which is the three-dimensional equivalent of the two-dimensional split-spread profile, sources/receivers are situated radially from a common receiver/source (Figure 30). The three-dimensional equivalents of the congruent rays in Figures 25 and 28 are those having equal angles of incidence to the normal of the interface at which reflection, or postcritical refraction occurs. In the three-dimensional case, however, these rays will not, in general, have the same suite of intra-layer angles, and it is by no means clear that they should have the same intercept times.

The rays in Figure 30 differ from the congruent rays of Figures 25 and 28, in that they do not necessarily have equal q's. However, they still have (nearly) identical intercept times. All 'congruent' rays in such a profile will have intercept times which vary only slightly, if at all. The intercept times are exactly equal for 'congruent' rays in a single layer and the 'congruent' rays of a reflecting ray I_1, I_2 will have $I_2 \cdot N = -I_1 \cdot N =$ constant. If $I_1 = \langle A_1, B_1, C_1 \rangle$ is the incident ray, the reflected (or critically refracted) ray, $I_2 = \langle A_1, B_2, C_2 \rangle$ will have a negative C_2, given by Equation (3). The intercept time for the two rays is then

$$\tau = uZ(C_1 - C_2) = uZ(C_1 - C_1 + 2\gamma I_1 \cdot N),$$

or

$$\tau = 2\Delta\tau(0)\,\gamma\cos(i),$$

where i is the reflection/refraction angle, $\Delta\tau(0)$ the two-way reflection time from the $X=Y=0$ origin to the layer at normal incidence, and γ is the direction

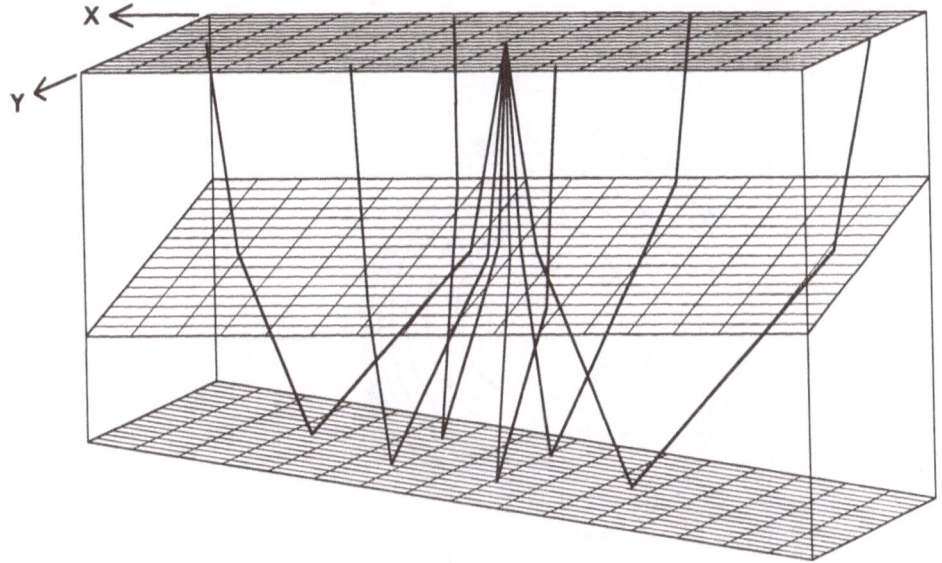

Fig. 30. As in the two-dimensional case (Figure 25), 'congruent' rays for a fixed-point profile in 3-D geometry. Each of these rays has the same angle of reflection with the bottom interface normal, and each has the same intercept time.

cosine of **N** with respect to vertical. Intercept time for reflection or headwave refraction in a single layer is constant for constant *i*. When multiple layers are involved, there will be small, systematic deviations in the intercept times of the 'congruent' rays, due to refraction across the overlying interfaces. These deviations are quite small; in the order of a few milliseconds for reasonable dips. In any case, the differences can be dealt with in the course of the inversion, as described below.

The τ's for the 'congruent' rays, then, resemble contours in the three-dimensional space $\tau(p_x, p_y)$ (Figure 31). These rays can be correlated with an ease and accuracy that depends on the quality of the $\tau(p_x, p_y)$ transformation.

POSTCRITICAL 3-D INVERSION

Just as in the case of 1-D $\tau(p)$ data, velocity-depth inversion can be performed using either postcritical or precritical arrivals. When X, Y offsets are large enough, the $\tau(p_x, p_y)$ or $\tau(p_R)$ data will contain postcritically reflected and refracted arrivals that can be easily picked, based on amplitude and position in the data field. For common-point data, the necessary 'congruent' τ's can be correlated by their equality. As in the 2-D case, it will, in general, be necessary to interpolate between the discretely sampled points to obtain the (p_x, p_y) [or, p_R] values for equal τ's.

Once the correct set of three or more rays with common postcritical reflection/ refraction angles from a particular interface have been identified, the determination of the orthogonal ray slownesses in the layer immediately above the reflecting/

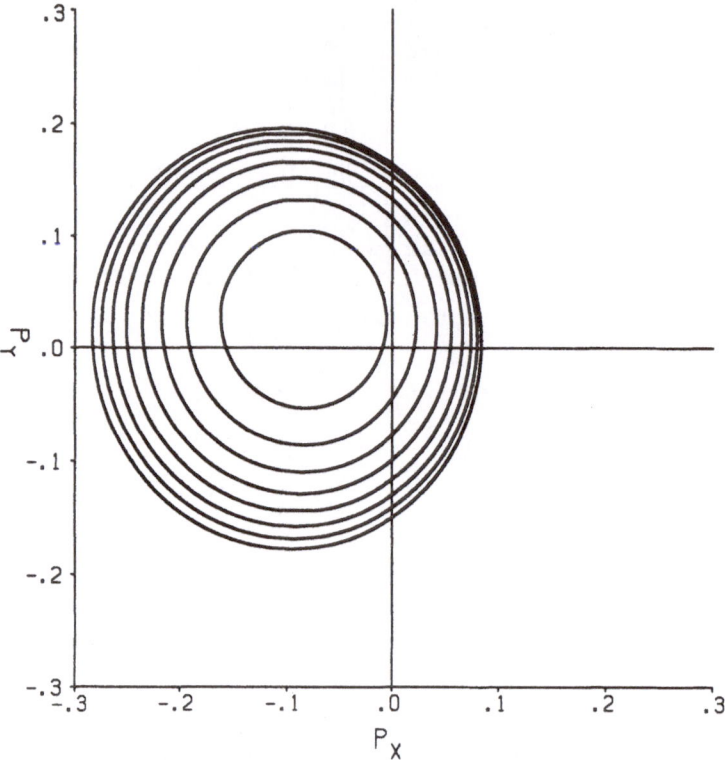

Fig. 31. In three dimensions, the constant τ's for 'congruent' rays in a fixed-point profile form intercept time contours. This figure shows contours of $\tau(p_x, p_y)$ for reflections from the second layer of the model of Figure 30.

refracting horizon can be accomplished in a manner analogous to the application of the two-dimensional algorithm described above. As before, the initiation of this process begins with ray angles.

In the case that $\tau(p_x, p_y)$ are in hand, these can be determined from the apparent slownesses p_x, p_y at the chosen points. Each ray is propagated downward through a known structure, using the 3-D refraction equation (2). When the layer of interest is reached, the three current values are used in the 3-D reflection/refraction equation (3) to solve for two of the direction cosines defining the interface normal and for the cosine of the ray/normal angle.

In the more common case that $\tau(p_R)$ has been obtained from data shot along various azimuths, the ray angles must be found as part of the solution, as well. If the three acquisition lines are written as vectors $\mathbf{T}_j = \langle X_j, Y_j, 0 \rangle$, then the observed radial slownesses are $p_{Rj} = \mathbf{P}_j \cdot \mathbf{T}_j$, where $\mathbf{P}_j$ is the slowness vector of the ray. Just as in the 2-D case, the problem can be divided into two parts; first the determination of velocity and dip, using only the observed horizontal slownesses, and second, the determination of layer thickness, using those values and the observed intercept times.

As described in the first section of this chapter, each interaction of a ray and an interface takes place in a plane defined by the ray and the layer normal. In a single layer, this plane also contains the track line, **T**. As shown in Figure 32, the angle between the track line and the layer normal, **N** in this plane is related to the apparent dip of the interface, as seen from the track line. Three equations can then be written, one for each of the observed p_R values:

$$\cos^{-1}(\mathbf{N}\cdot\mathbf{I}_j) + D_j = 180 - \cos^{-1}(\mathbf{T}_j\cdot\mathbf{I}_j), \quad j = 1,2,3.$$

Here, the known values are on the right-hand side, with $p_{Rj} = u_1(\mathbf{T}_j\cdot\mathbf{I}_j)$, and the unknown values on the left. At first glance, it would appear that there are four unknowns: the constant ray-layer normal angle $\cos^{-1}(\mathbf{N}\cdot\mathbf{I}_j)$, and the three apparent dip angles D_j, but only two of the D_j's are independent. Therefore, the three equations can be solved to determine the apparent dips (and, equivalently, **N**) and the critical angle $\cos^{-1}(\mathbf{N}\cdot\mathbf{I}_j)$. This procedure is equivalent to that presented by Russell *et al.* (1982) but is presented here in more general terms.

Since the positions and amplitudes of congruent $\tau(p)$ picks have been used to identify them as postcritical arrivals, the ray/normal angle, combined with the slowness previously determined for the layer in which the reflection/refraction occurs, gives the slowness of the layer immediately below the interface. With this value in hand, the q's may be calculated for the three rays. The q's are then

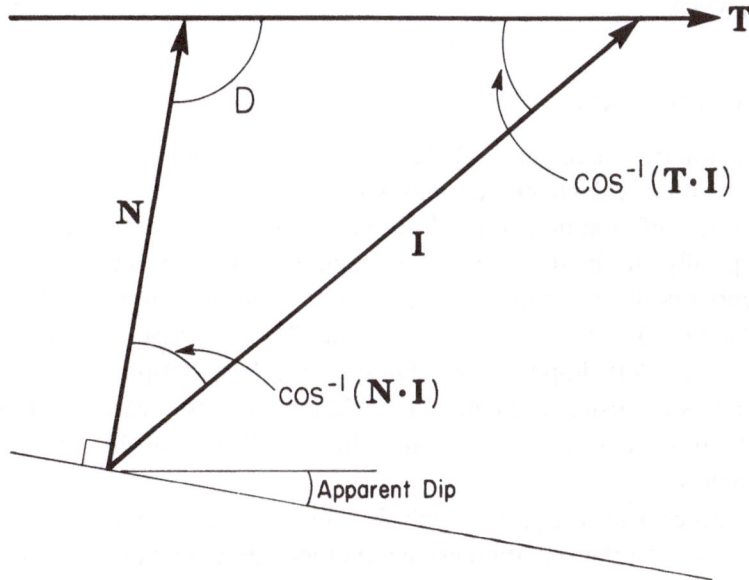

Fig. 32. In a single layer above a three-dimensionally dipping interface, a reflected or refracted ray (**I**), the layer normal (**N**) and the source-receiver track (**T**) will lie in a common plane. The angle between the ray and the track is determined from the apparent slowness (p_R) of the ray along the track. Observations of slowness of rays with the same refraction/reflection angle $\cos^{-1}(\mathbf{N}\,\mathbf{I})$ along three different tracks are sufficient to determine the angle, as well as the strike and dip of the interface.

combined with the observed intercept times to give the thickness of the overlying layer, using the Tau-sum inversion of Equation (13). As with the 1-D and 2-D methods, it is necessary to perform the inversion for layers in each iteration. In the case that the input data are $\tau(p_R)$ values, the downward continuation, or layer stripping, is not quite as straightforward as it is when $\tau(p_x, p_y)$ data are in hand. A single p_R observation is sufficient only to define a cone, whose axis is the track line, and on whose surface the ray must lie. The ray tracing equations (2) can be used to propagate any given ray downwards, but the exact ray angle will only be determined when the critical layer is reached. Therefore, the ray bending factors of Equation (2) must be carried along in the analysis. Fortunately, this procedure is easy enough to perform in computer software.

While in many cases the small deviations of congruent τ's from equality will not induce great errors, the presence of such errors can be easily detected, by comparing the layer thicknesses calculated using each of the three $\tau(p_x, p_y)$ [or, $\tau(p_R)$] values, and the three sets of q_{aj}, q_{bj} values independently. When the differences are significant, an iteration can be performed to correct for the refraction-induced errors. The current inversion solution for the layer normals and velocities are used in Equations (2) and (3) to predict the slight differences in the appropriate τ times for two of the three values used for each layer. These are re-picked, and the inversion is performed again. Then, the layer thickness determinations using each of the three sets of q's are compared, to see if further iteration is required.

PRECRITICAL 3-D INVERSION

The greatest amount of existing 3-D data, by far, has been acquired as CDP reflection data with areal coverage. Although these data can, in general, be regathered in terms of common source/receiver points at the track intersections, the offsets typically attained are small enough to preclude the recording of appreciable amounts of postcritical energy. These data may, however, be easily transformed to $\tau(p_R)$, or in the case of true 3-D coverage, to $\tau(p_x, p_y)$. Dealing with $\tau(p_R)$ data imposes an additional level of complexity, just as it does in the analysis of postcritical data. These added difficulties can be handled in ways similar to those described above and, for simplicity, we discuss only the 'simple' case below.

Analysis of precritical $\tau(p_x, p_y)$ arrivals can be carried out in a manner somewhat analogous to the one-dimensional methods described above. When the intercept times for passage through a single layer are isolated, they are described by

$$\Delta\tau(p_x, p_y) = \Delta\tau(0)\,\gamma\cos(i), \tag{14}$$

where i is the angle of incidence of the ray with respect to the layer normal. If the maximum $\Delta\tau(p_x\max, p_y\max) = \Delta\tau(0)\gamma$ can be located, the layer normal is simply

$$\mathbf{N} = \langle\alpha, \beta, \gamma\rangle = \langle p_x\max, p_y\max, q\max\rangle/u,$$

where

$$q\max = (1 - p_x\max^2 - p_y\max^2)^{1/2},$$

and the layer slowness can be obtained from any other $\Delta\tau(p_x, p_y)$ observation, using Equation (14) and the observation that

$$\cos(i) = \Delta\tau(p_x, p_y)/\Delta\tau(0)\gamma = \mathbf{I}\cdot\mathbf{N} = (p_x\alpha + p_y\beta + q\gamma). \qquad (15)$$

Therefore,

$$u = \Delta\tau(0)\gamma[p_x\alpha + p_y\beta + (1 - p_x^2 - p_y^2)^{1/2}]/\Delta\tau(p_x, p_y),$$

and the layer thickness is given by $2Z = \Delta\tau(0)/u$. If maximum $\Delta\tau = \Delta\tau(0)\gamma$ is not directly observed, the p_x, p_y values from three identical τ's can be used for the simultaneous solution of Equation (15) for $\mathbf{N}$, u, and Z.

Once these values have been determined for the uppermost layer, its $\Delta\tau(p_x, p_y)$ contributions to the τ's of underlying layers can be analytically calculated using Equation (15) and subtracted, in a manner analogous to Schultz's (1982) method. In the 3-D case, however, it is also necessary to apply a 2-D shift to the p_x, p_y values, according to Equation (2). This need only be done to compensate for the refraction of the down-going rays (when the receiver is at the common midpoint), but it must be accomplished by making an analytic correction during solution for $\mathbf{N}$, u, and $\Delta\tau(0)$ for subsequent layers. The inversion for the first layer yields the slowness of the layer (u_1) and the normal vector ($\mathbf{N}_1$) of the interface below. According to Equation (2), the shifted values $\langle p_x, p_y\rangle_2$ are given by

$$\langle p_x, p_y\rangle_2 = \langle p_x, p_y\rangle_1 - \mathbf{N}_1[(u_2^2 - u_1^2 - (\mathbf{S}_1\cdot\mathbf{N}_1)_2)^{1/2} - (\mathbf{S}_1\cdot\mathbf{N}_1)],$$

where $\mathbf{S}_1 = \langle p_x, p_y, q\rangle_1$. u_2, the slowness in the next layer, is as yet unknown, so this correction must be carried along during the solution for u_2. If this layer stripping step is not taken, the analysis can still be carried out, in which case, the results will be exactly equivalent to the time domain treatment described by Levin (1971), and will suffer the usual limitations on the recovery of interval velocities with methods based on the Dix equation.

Anisotropy

The presence of anisotropy tends to complicate traveltime interpretation. The inclusion of anisotropic effects in the use of Equation (1) was described by Hake (1986). In the very special case that the anisotropy is elliptical, a straightforward approach can be taken. For transverse isotropy, the medium slowness can be described by its vertical and horizontal components. In this case, observations of horizontal ray slowness can only give information about the horizontal component of the medium's slowness, and the corresponding intercept times depend only on the vertical slowness. Data from vertical arrays, however, will contain the necessary additional information about the vertical slowness of the medium. In a two-dimensional model with transverse isotropy, four independent determinations will be required to resolve the parameters for a single layer. With three-dimensionally elliptical anisotropic layers separated by three-dimensionally dipping interfaces, there will be six unknown parameters defining each layer, and a corresponding number of independent observations required to resolve them.

In a generally anisotropic medium, the angular dependency of slowness will not be so simple as in the elliptic case. Under some simplifying assumptions about the nature of the anisotropy, (cf. Radovich and Levin, 1982), some interpretation of reflected and refracted rays may still be possible within the simple framework presented here. Anisotropic parameters will not always be so simple, however (Thomsen, 1986), and the presence of dip further complicates the interpretation (Krey and Helbig, 1956).

Acknowledgements

The research leading to the results shown here was funded in part by the following U.S. Government contracts and grants: NSF; OCE76-82328, OCE79-22884, OCE83-19518, DPP85-17635 ONR; N00014-80-C-0098, N00014-84-C-0132. Lamont-Doherty contribution #4474.

References

Adachi, R., 1954, On a proof of fundamental formula concerning refraction method of geophysical prospecting and some remarks, *Kumamoto J. Sci.*, *Ser. A* **2**, 18–23.
Al-Chalabi, M., 1974, An analysis of stacking, RMS, average, and interval velocities over a horizontally layered ground, *Geophys. Prosp.* **22**, 458–475.
Bessonova, E. N., Fishman, V. M., Ryaboyi, V. Z., and Sitnikova, G. A., 1974, The tau method for inversion of travel times — I. Deep seismic sounding data, *Geophys. J. Roy. Astr. Soc.* **36**, 377–398.
Brocher, T. M. and Phinney, R. A., 1981, Inversion of slant stacks using finite-length record sections, *J. Geophys. Res.* **86**, 7065–7072.
Cutler, R. T. and Love, P. L., 1980, Elliptical velocity analysis [abstract], *Geophysics* **45**, 540.
Diebold, J. B. and Stoffa, P. L., 1981, The traveltime equation, tau-p mapping and inversion of common midpoint data, *Geophysics* **46**, 238–254.
Diebold, J. B., Stoffa, P. L., Buhl, P., and Truchan, M., 1981, Venezuela Basin crustal structure, *J. Geophys. Res.* **86**, 7901–7923.
Diebold, J. B., 1987, Three dimensional traveltime equation for dipping layers, *Geophysics* **52**, 1492–1500.
Dix, C. W., 1955, Seismic velocities from surface measurements, *Geophysics* **20**, 68–86.
Ewing, J. I., 1963, Elementary theory of seismic refraction and reflection measurements, in M. N. Hill (ed.), *The Sea*, Vol. 3, John Wiley, New York.
Ewing, M., Woolard, G. P., and Vine, A. C., 1939, Geophysical investigation in the emerged and submerged Atlantic coast plain, Part III: Barnegat, New Jersey section, *GSA Bull.* **50**, 257–296.

Garmany, J., Orcutt, J. A., and Parker, R. L., 1979, Travel time inversion: a geometrical approach, *J. Geophys. Res.* **84**, 3615–3622.

Greenhalgh, S. A. and King, D. W., 1981, Curved raypath interpretation of seismic refraction data *Geophys. Prosp.* **29**, 853–882.

Hake, H., 1986, Slant stacking and its significance for anisotropy, *Geophys. Prosp.* **34**, 595–608.

Johnson, L. R. and Lee, R. C., 1985, Extremal bounds on the *P* velocity in the earth's core, *SSA Bull.* **75**, 115–130.

Johnson, S. H., 1976, Interpretation of split-spread refraction data in terms of plane dipping layers, *Geophysics* **41**, 418–424.

Kaufman, H., 1953, Velocity functions in seismic prospecting, *Geophysics* **18**, 289–299.

Kennett, B. L. N., 1976, A comparison of travel time inversions, *Geophys. J. Roy. Astr. Soc.* **44**, 517–536.

Krey, T. and Helbig, K., 1956, A theorem concerning anisotropy of stratified media and its significance for reflection seismics, *Geophys. Prosp.* **34**, 294–302.

Levin, F. K., 1971, Apparent velocity from dipping interface reflections, *Geophysics* **36**, 510–516.

Milkereit, B., Mooney, W. D., and Kohler, W. M., 1985, Inversion of seismic refraction data in planar dipping structures, *Geophys. J. Roy. Astr. Soc.* **82**, 81–103.

Ocola, L., 1972, A nonlinear least-squares method for seismic refraction mapping — Part II: Model studies, Appendix, *Geophysics* **37**, 273–287.

Phinney, R. A., Roy Chowdhury, K., and Frazer, L. N., 1981, Transformation and analysis of record sections, *J. Geophys. Res.* **86**, 359–377.

Radovich, B. J. and Levin, F. K., 1982, Instantaneous velocities and reflection times for transversely isotropic solids, *Geophysics* **47**, 316–322.

Russell, D. R., Keller, G. R., and Braile, L. W., 1982, A technique to determine the three-dimensional attitude and true velocity of a refractor, *Geophysics* **47**, 1331–1334.

Schultz, P. S., 1976, Velocity estimation by wave front synthesis, PhD thesis, Stanford University, Stanford, Ca.

Schultz, P. S., 1982, A method for direct estimation of interval velocities, *Geophysics* **47**, 1657–1671.

Schultz, P. S., Pierprzak, A. W., and Loh, I. K. L., 1983, A case for larger offsets, *Geophysics* **48**, 238–247.

Slotnick, M. M., 1936, On seismic computations with applications, I, *Geophysics* **1**, 9–22.

Stark, P. B., Parker, R. L., Masters, G., and Orcutt, J. A., 1986, Strict bounds on seismic velocity in the spherical earth, *J. Geophys. Res.* **91**, 13,892–13,902.

Stoffa, P. L., Diebold, J. B., and Buhl, P., 1981, Inversion of data in the Tau-*p* plane, *Geophys. Res. Lett.* **8**, 869–872.

Stoffa, P. L., Diebold, J. B., and Buhl, P., 1982, Velocity analysis for wide aperture seismic data, *Geophys. Prosp.* **30**, 25–57.

Thomsen, L., 1986, Weak elastic anisotropy, *Geophysics* **51**, 1954–1966.

Treitel, S., Gutowski, P. R., and Wagner, D. E., 1982, Plane-wave decomposition of seismograms, *Geophysics* **47**, 1375–1401.

Vera E. and Diebold, J. B., 1984, Generalized inversions of discrete Tau-*p* data, *EOS* **65**, 239.

Vera, E., 1987, On the connection between the Herglotz–Weichert–Bateman and Tau-sum inversions, *Geophysics* **52**, 568–570.

Lamont-Doherty Geological
  Observatory,
Rt. 9W,
Palisades,
NY 10964, U.S.A.

M. TYGEL, P. HUBRAL and F. WENZEL

Plane-Wave Decomposition: A Tool for Deconvolution

Abstract

Most (prestack) deconvolution methods for seismic shot records are based on simple stochastic models for the traces. They do not exploit the fact that these traces have to satisfy the wave equation. *A plane-wave decomposition* of a point-source seismogram recorded over a vertically inhomogeneous layered (acoustic) medium is a process based on the wave equation. It offers the possibility to extract both (a) the causal source pulse of arbitrary unknown shape and (b) the unknown broad-band plane-wave response (reflectivity) for any incidence angle or ray parameter. This was hitherto considered to be impossible but has become a reality if one exploits the wave theory that permits us to compose and decompose point-source responses from vertically inhomogeneous media in terms of plane waves. As starting point for constructing the seismogram by a *plane-wave composition* serves the *Weyl Integral* in either its time-harmonic or transient form. The central concept, on the other hand, upon which the *plane-wave decomposition* is based is the *slant stack*. The theory of plane-wave composition is briefly reviewed in order to establish the background for the new deconvolution procedure which essentially uses the properties that the reflectivity function (i.e. the reflection response of a layered medium for an incident plane wave) has for above-critical incidence angles.

1. Introduction

The exploration seismologist is confronted with two fundamental problems in connection with extracting the physical subsurface parameters from a point-source seismogram recorded over a horizontally stratified medium. These are the problems of *deconvolution* (i.e., the recovery of the source pulse) and *inversion* (i.e., the recovery of the impedance-depth profile).

For the point-source response of a vertically layered isotropic (acoustic) medium, it is well known (see, e.g., Müller, 1971) that the theory of plane-wave decomposition (PWD) constitutes an important step towards solving the inversion problem. Once the point-source seismogram (PSS) is deconvolved with the extracted source pulse (i.e., after the Green function is established), it can be decomposed into broad-band plane-wave seismogram (PWS). These can, in

Paul L. Stoffa (ed.), Tau-p: A Plane Wave Approach to the Analysis of Seismic Data, 119—139.
© *1989 by Kluwer Academic Publishers.*

principle, be inverted to construct the desired impedance-depth profile (Bube and Burridge, 1983; Sonnevend, 1987; Symes, 1983; Ursin and Berteusen, 1986; Yagle and Levi, 1984).

Though the theory of inverting a PSS, via a PWD as an intermediate step, is in itself mathematically elegant and appealing, its practical realization encounters many obstacles. One possible source for errors provides the often incomplete deconvolution due to the frequently unknown source pulse. It is this topic that we will give special attention to in this chapter.

Exploration seismologists live in general with the conviction that to construct the unknown reflectivity function (broad-band plane-wave reflection response) from a seismogram, one must know the source wavelet or, vice-versa, in order to extract the source wavelet one must know the reflectivity.

In other words, it is widely accepted that one cannot extract both the reflectivity and the unknown (mixed-delay) wavelet as long as no restrictions are made on both — e.g., that the reflectivity function has a white spectrum and the source wavelet is minimum delay (Robinson, 1954).

In this chapter we describe a theory whereby both the reflectivity and the wavelet can be recovered from the point-source response of a layered vertically inhomogeneous medium. No specific restrictions have to be put on either the wavelet or the distribution of reflection coefficients which determine the reflectivity function. The only assumption that is to be made is that the recorded seismic traces represent the response of a point source from a vertically inhomogeneous medium. More precisely, all what is assumed is that the data that are to be analyzed result from a wave field that originates in a point source and satisfies the appropriate wave equation in each layer as well as the usual boundary conditions at the interfaces. Such assumptions were also made by Fokkema and Ziolkowski (1985), Ziolkowski et al. (1987) and Kelamis et al. (1987) in their attempt to deconvolve a shot-record with the help of the critical reflection theorem, which will be reviewed below. While the critical reflection theorem permits the determination of the amplitude spectrum of the unknown source pulse (and also the wavelet itself provided that there is minimum delay) one needs one more theorem to also extract the phase spectrum of the unknown pulse. This theorem will be presented in this work.

To keep our analysis simple, we assume below the model of Figure 1. A spherical source is located at point S in the upper half-space above the layered medium. The medium consists of homogeneous acoustic isotropic layers, designated by the density-velocity pairs $(p_j, c_j = 1/p_j, j = 0, \ldots, N+1)$ and the depth values H_j $(j = 0, \ldots, N)$. The quantity p_j is the *medium slowness* of layer j.

The source itself, placed at point S $(r = 0, z = -h)$, is defined by the acoustic potential field

$$\phi_i(r, z, t) = f(t-R/c_0)/R, \quad (f(t) = 0 \quad \text{for} \quad t < 0), \tag{1}$$
$$R = \sqrt{r^2 + (z + h)^2},$$

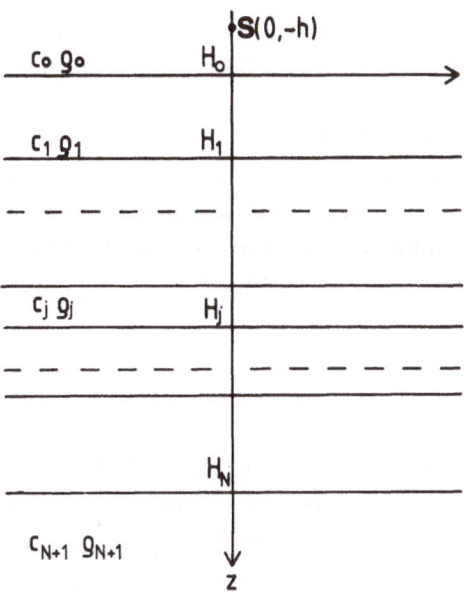

Fig. 1. Model used for the calculation.

whereby the function $f(t)$ describes an arbitrary unknown wavelet. We will operate with acoustic potentials. It is to be recalled (Brekhovskikh, 1980) that the pressure field is simply obtained from the acoustic potential field by a multiplication by the medium density.

In the following, we show how the unknown function $f(t)$ can be extracted from the PSS, which is assumed to be recorded along a profile in the upper half-space $z = -h^*$, $(0 < r < \infty)$. There is no particular assumption made with respect to the layered medium.

Just as much as a plane-wave decomposition (PWD) provides the key to extracting plane-wave responses from point-source responses, it also provides the key for the determination of $f(t)$. In fact, both processes are closely related, as we will see.

The theory of PWD of a point-source reflections response will now be briefly reviewed to prepare the reader for the extraction of $f(t)$ out of the PSS. To get a better understanding of the theory of PWD, we also summarize the theory of plane-wave composition (PWC). We address both the time-harmonic and transient situations.

The most convenient starting point for introducing the reader to the theory of PWC of a point-source seismogram is, in our opinion, the Weyl integral (Weyl, 1919). It will be briefly reviewed in both its time-harmonic and transient versions. Thereafter it will be shown how PSSs are constructed with the help of the Weyl integral in either form.

Just as much as the Weyl integral signifies the central concept to formulate the

PSS via a PWC, the so-called slant-stack denotes the central concept upon which the theory of PWD can be based. The PWD of a PSS is closely connected to the Radon transform.

This chapter is structured as follows. In section 2 we summarize the PWC theory and in Section 3 the PWD theory. In the first part of each Section we treat the time-harmonic theory and in the second part, the transient theory. In Section 4, we discuss the new deconvolution method. The formal steps leading to the extraction of $f(t)$ can be easily conceived and are therefore only summarized below.

1.1. SUMMARY OF DECONVOLUTION PROCEDURE

Figure 2 shows a sketch of the traveltime curves associated with the reflection pressure seismogram $\rho_0 \phi_S(r, z = -h^*, t)$ for the source pulse (1). Also indicated is the ensemble of lines $t = \tau + rp$ $(-\infty < \tau < \infty, 0 < r < \infty, p = \text{const.})$, where τ is the intercept time and p is the ray parameter.

As shown below, the PSS ϕ_S, namely the superposition of the source potential ϕ_i and its reflection response ϕ_r at offset $r > 0$, level $z = -h^*$ and time t, can be constructed, via a PWC, to provide

$$\phi_S(r, z = -h^*, t) = \phi_i(r, z = -h^*, t) + \phi_r(r, z = -h^*, t) \tag{2a}$$

$$= \text{Re}\left\{ \frac{1}{\pi} \int_0^\infty d\omega \, e^{i\omega t} \, \hat{\phi}_S(r, z = -h^*, \omega) \right\} \tag{2b}$$

$$= \text{Re}\left\{ \frac{1}{\pi} \int_0^\infty d\omega \, e^{i\omega t} [\hat{\phi}_i(r, z = -h^*, \omega) + \right.$$

$$\left. + \hat{\phi}_r(r, z = -h^*, \omega)] \right\}, \tag{2c}$$

where

$$\hat{\phi}_i(r, z = -h^*, \omega)$$

$$= \hat{f}(\omega)(-i\omega) \int_0^\infty dp \, \frac{p}{P_0(p)} J_0(rp\omega) \, e^{-i\omega|h-h^*| P_0(p)} \tag{2d}$$

and

$$\hat{\phi}_r(r, z = -h^*, \omega)$$

$$= \hat{f}(\omega)(-i\omega) \int_0^\infty dp \, \frac{p}{P_0(p)} J_0(rp\omega) \, \hat{R}(p, \omega) \, e^{-i\omega(h+h^*) P_0(p)}, \tag{2e}$$

in which $J_0(\alpha)$ denotes the Bessel function of order 0, $\hat{f}(\omega)$ is the Fourier spectrum of $f(t)$, and $\hat{R}(p, \omega)$ is the (time-harmonic) reflectivity function. $P_0(p)$ is defined in Equation (8d).

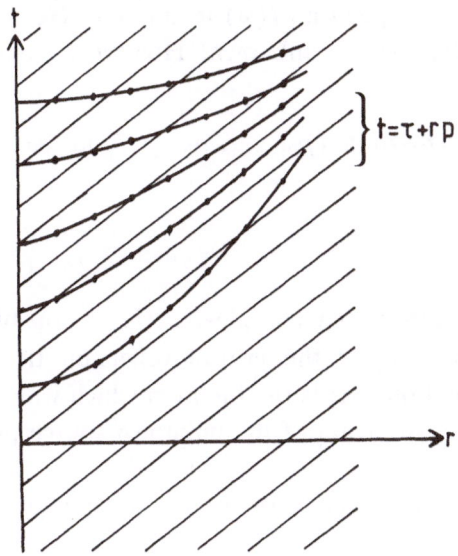

Fig. 2.   Sketch of seismic reflections and linear trajectories used for the stack.

From the above equations, we can readily write the following expression for the Fourier spectrum: $\hat{\phi}_S(\omega)$ of the PSS:

$$\hat{\phi}_S(r, z = -h^*, \omega)$$

$$= \hat{f}(\omega)(-i\omega) \int_0^\infty \frac{dpp}{P_0(p)} J_0(rp\omega) \left[e^{-i\omega|h-h^*|P_0(p)} + \right.$$

$$\left. + \hat{R}(p, \omega) e^{-i\omega(h+h^*)P_0(p)} \right]. \tag{2f}$$

Let us now recall the definition of the Fourier transform of $\phi_S(r, z = -h^*, t)$

$$\hat{\phi}_S(r, z = -h^*, \omega) = \int_{-\infty}^\infty dt \, e^{-i\omega t} \, \phi_S(r, z = -h^*, t) \tag{3a}$$

to introduce, for any given ray parameter $p > 0$ and any frequency $\omega > 0$, the new data function

$$\hat{G}_S(p, \omega) = \int_0^\infty dr \, r \, J_0(rp\omega) \, \hat{\phi}_S(r, z = -h^*, \omega). \tag{3b}$$

Note that $\hat{G}_S(p, \omega)$ is directly obtained from the recorded PSS $\phi_S(r, z = -h^*, t)$ and may be considered, in principle, for each $p > 0$ and $\omega > 0$, as known data.

Using well-known properties of the so-called Hankel or Fourier–Bessel transformations of order 0, we find, from Equations (2f) and (3b), the fundamental relationship

$$\left(-\frac{i}{\omega}\right) \frac{\hat{f}(\omega)}{P_0(p)} \left[e^{-i\omega|h-h^*|P_0(p)} + \hat{R}(p, \omega) e^{-i\omega(h+h^*)P_0(p)} \right] = \hat{G}_S(p, \omega). \tag{4}$$

The important Equation (4) relates the spectrum $\hat{f}(\omega)$ to the data $\hat{G}_S(p, \omega)$. Note that the reflectivity function $\hat{R}(p, \omega)$ is unknown. However, from the properties (see Appendix)

$$\hat{R}(p, \omega) \text{ real}; \quad |\hat{R}(p, \omega)| < 1 \quad \text{(if } p > p_M = \max(p_0, \ldots, p_{N+1}) \quad \text{and} \quad \omega > 0)$$
$$(5a)$$

and

$$|\hat{R}(p, \omega)| = 1 \quad \text{(if } p_{N+1} < p < p_0),$$
$$(5b)$$

we show in Section 4 the following result (we consider, as is reasonable in practice, that $p_{N+1} < p_0$). Condition (5b) is the critical reflection theorem (Fokkema and Ziolkovski, 1985). Condition (5a) is the theorem which will make possible the determination of the phase spectrum of the unknown source pulse.

MAIN RESULT: Let $\tilde{p}$ and $\tilde{\tilde{p}}$ be any given values of the ray parameter, satisfying the conditions

$$\tilde{p} > p_M = \max(p_0, \ldots, p_{N+1})$$
$$(6a)$$

and

$$p_{N+1} < \tilde{\tilde{p}} < p_0.$$
$$(6b)$$

Then, for all $\omega > 0$, we have

$$\arg\{\hat{f}(\omega)\} = \arg\{\hat{G}_S(\tilde{p}, \omega)\}$$
$$(6c)$$

and

$$|\hat{f}(\omega)| = \frac{\omega |P_0(\tilde{\tilde{p}})| |G_S(\tilde{\tilde{p}}, \omega)|}{2 \cos \alpha(\tilde{\tilde{p}}, \omega)},$$
$$(6d)$$

where

$$\alpha(\tilde{\tilde{p}}, \omega) = \arg\{\hat{G}_S(\tilde{\tilde{p}}, \omega)\} + \arg\{\hat{f}(\omega)\} + \pi/2 - \omega|h - h^*|P_0(\tilde{\tilde{p}}). \quad (6e)$$

Equations (6c–d) define $\hat{f}(\omega)$, namely

$$\hat{f}(\omega) = |\hat{f}(\omega)| \exp[i \arg\{\hat{f}(\omega)\}]$$
$$(6f)$$

for all $\omega > 0$. In summary, the above result expresses that, in principle, only two values p and p satisfying the easy conditions (6a–b) completely specify $\hat{f}(\omega)$ for all $\omega > 0$.

Hence, we can construct $f(t)$ by

$$f(t) = \text{Re}\left\{\frac{1}{\pi} \int_0^\infty d\omega \, e^{i\omega t} \hat{f}(\omega)\right\}.$$
$$(6g)$$

Next, the reflectivity function $R(p, t)$ is given, likewise, by

$$R(p, t) = \text{Re}\left\{\frac{1}{\pi} \int_0^\infty d\omega \, e^{i\omega t} \hat{R}(p, \omega)\right\},$$
$$(6h)$$

where $\hat{R}(p, \omega)$ is obtained from Equation (4) using that $\hat{f}(\omega)$ is now known. We have for any value of p

$$\hat{R}(p, \omega) = e^{i\omega(h+h^*)\,P_0(p)} \left\{ \frac{\omega P_0(p)}{-i\hat{f}(\omega)} \hat{G}_S(p, \omega) - e^{-i\omega|h-h^*|\,P_0(p)} \right\}. \tag{6i}$$

This completes the recovery of $f(t)$ and $R(p, t)$ as claimed.

2. Plane-Wave Composition

The function $f(t - R/c_0)/R$ in Equation (1) can be written as a superposition of time-harmonic point sources

$$f(t - R/c_0)/R = \mathrm{Re}\left\{ \frac{1}{\pi} \int_0^\infty d\omega\, \hat{f}(\omega) \frac{e^{i\omega(t-R/c_0)}}{R} \right\}. \tag{7}$$

We show below how this transient acoustic potential can be further decomposed (say, for observations points below the source level) into plane waves, which radiate away from the source point S. In the upper half-space, each plane-wave component will be accompanied by a corresponding reflected plane wave, so that the PSS can be thought of as a composition (or simulation) of reflected plane waves (the reflection response) plus the incident source potential.

We begin by reviewing the composition of the time-harmonic point source $\exp[i\omega(t-R/c_0)]/R$ in terms of time-harmonic plane waves. Thereafter, we show how the transient point source potential (1) can also be directly simulated in terms of transient plane waves. Finally, we formulate the time-harmonic, as well as the transient reflection response of the layered medium of Figure 1, in terms of time-harmonic and transient plane-waves, respectively.

2.1. TIME-HARMONIC WEYL AND SOMMERFELD INTEGRALS

The time-harmonic point source can be expressed as a superposition of time-harmonic plane waves by the classical Weyl integral (see, e.g., Tygel and Hubral, 1987, Ch. 2) as

$$\frac{1}{R} \exp[i\omega(t - R/c_0)]$$

$$= -\frac{i\omega}{2\pi} \int_0^\infty dp \frac{p}{P_0(p)} \int_0^{2\pi} d\varphi \exp[i\omega(t - \eta p - (z + h)P_0(p))], \tag{8a}$$

$$(z + h \geqslant 0),$$

where p is the ray parameter or horizontal slowness,

$$\eta = x \cos \varphi + y \sin \varphi, \tag{8b}$$

$$R = \sqrt{r^2 + (z+h)^2}; \quad r = \sqrt{x^2 + y^2} \tag{8c}$$

and

$$P_0(p) = \sqrt{1/c_0^2 - p^2} = \begin{cases} |P_0(p)| & \text{if } 0 \leqslant p \leqslant 1/c_0, \\ -i|P_0(p)| & \text{if } p > 1/c_0. \end{cases} \quad (8d)$$

As indicated in Equation (8a), the above representation is valid only for $z+h \geqslant 0$, i.e., for observation points $P(r, z)$ below the source level. In order to have a valid representation in full space, we must introduce the absolute value $|h+z|$ to replace $z+h$.

For $0 \leqslant p \leqslant 1/c_0$ (homogeneous waves), the plane-wave components in Equation (8a) propagate in the direction $n = (\cos \varphi \vec{i} + \sin \varphi \vec{j}) \sin \theta + \cos \theta \vec{k}$ with $p = \sin \theta/c_0$. Figure 3 shows the geometric significance of these angles and vectors. Amplitudes are constant on these wavefronts. For $p > 1/c_0$ (inhomogeneous waves), they propagate horizontally exhibiting vertical wavefronts with exponentially decaying amplitudes in the positive z direction.

The Weyl integral (8a) may be easily changed into the so-called Sommerfeld integral by performing the φ-integration. We obtain

$$\exp[i\omega(t - R/c_0)]/R$$
$$= -i\omega \int_0^\infty dp \frac{p}{P_0(p)} J_0(rp\omega) \exp[i\omega(t - (z + h)P_0(p))] \quad (9)$$
$$(r > 0; z + h \geqslant 0).$$

2.2. TRANSIENT WEYL AND SOMMERFELD INTEGRALS

Integration over positive frequency ($\omega > 0$) in Equation (8a) provides the so-called transient Weyl integral (Poritzky, 1951; Tygel and Hubral, 1987, Chapters 2 and 8)

$$\Delta(t-R/c_0)/R$$
$$= \frac{d}{dt}\left\{ -\frac{1}{2\pi} \int_0^\infty dp \frac{p}{P_0(p)} \int_0^{2\pi} d\varphi \, \Delta(t - \eta p - (z+h)P_0(p)) \right\},$$
$$(z+h \geqslant 0), \quad (10a)$$

where for complex ξ with $\text{Im } \xi \geqslant 0$,

$$\Delta(\xi) = \begin{cases} \delta(\xi) + i/\pi\xi & (\text{real } \xi), \\ i/\pi\xi & (\text{Im } \xi > 0) \end{cases} \quad (10b)$$

is called the Δ-function. This function is analytic in the upper half-plane $\text{Im } \xi > 0$ and is such that its real part coincides with the usual (real) Dirac δ-function on the real ξ-axis. The Δ-function $\Delta(\xi)$ ($\text{Im } \xi \geqslant 0$) may be seen in transient propagation to play the same role as the exponential function $\exp[i\omega\xi]$ in time-harmonic propagation. One can also say that $\Delta(\xi)$ ($\text{Im } \xi \geqslant 0$) is the analytic signal associated with the real signal $\delta(t)$. A systematic study of the Δ-function,

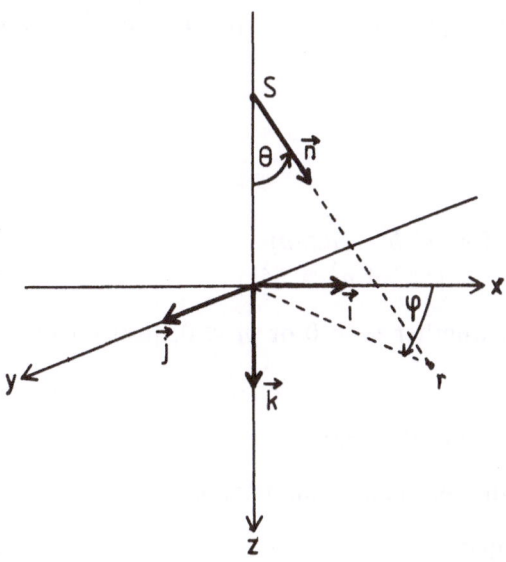

Fig. 3.   Normal vector pointing away from the source center in direction $\vec{n}$.

as well as its applications to transient propagation in stratified media, is done in Tygel and Hubral (1987).

For the arbitrary transient point-source potential (1), one can, likewise, write

$$f(t - R/c_0)/R$$

$$= f'(t) * \mathrm{Re}\left\{-\frac{1}{2\pi}\int_0^\infty \mathrm{d}p\, \frac{p}{P_0(p)}\int_0^{2\pi}\mathrm{d}\varphi\, \Delta(t - \eta p - (z+h)P_0(p))\right\}$$

$$(z+h \geq 0),\tag{10c}$$

where the prime denotes the derivative with respect to the argument and $*$ denotes convolution over time t.

The integral over φ in the above formulas (10a) and (10c) can be performed (Tygel and Hubral, 1987, Ch. 2) to yield the so-called transient Sommerfeld integral (for positive offset $r > 0$)

$$\delta(t - R/c_0)/R$$

$$= \frac{\mathrm{d}}{\mathrm{d}t}\mathrm{Re}\left\{-\frac{1}{\pi}\int_0^\infty \mathrm{d}p\, \frac{p}{P_0(p)}[r^2 p^2 - (t - (z+h)P_0(p))^2]^{-1/2}\right\}$$

$$(r > 0;\; z+h \geq 0),\tag{11a}$$

or, more generally,

$$f(t - R/c_0)/R$$

$$= f'(t) * \mathrm{Re}\left\{-\frac{1}{\pi}\int_0^\infty \mathrm{d}p\, \frac{p}{P_0(p)}[r^2 p^2 - (t - (z+h)P_0(p))^2]^{-1/2}\right\}$$

$$(r > 0;\; z+h \geq 0).\tag{11b}$$

In the above functions, the complex square root is of type $[b^2 - a^2]^{1/2}$ with $b > 0$ and $\mathrm{Im}\, a \geq 0$. This root is defined as

$$\mathrm{Re}\{\sqrt{b^2 - a^2}\} > 0 \quad \text{if } \mathrm{Im}\, a > 0 \tag{12a}$$

and

$$\sqrt{b^2 - a^2} = \begin{cases} |b^2 - a^2|^{1/2} & \text{if } a^2 < b^2 \quad (\text{rea}/a) \\ -i\,\mathrm{sgn}(a)\,|b^2 - a^2|^{1/2} & \text{if } a^2 > b^2 \end{cases} \tag{12b}$$

where $\mathrm{sgn}(m) = \pm 1$ according to whether $m \geq 0$ or $m < 0$, respectively.

2.3. TIME-HARMONIC POINT-SOURCE SEISMOGRAM

As explained in Section 2.1, the time-harmonic point-source potential

$$\Phi_i^\omega = \hat{f}(\omega) \exp[i\omega(t - R/c_0)]/R \tag{13a}$$

can be synthesized from the incident time-harmonic plane waves

$$\Phi_I^\omega = -i\omega \hat{f}(\omega) \exp[i\omega(t - \eta p - (z+h)P_0(p))], \quad (z+h \geq 0). \tag{13b}$$

Each of the above waves possesses in the upper half-space $z < 0$, a reflected companion plane-wave

$$\Phi_R^\omega = -i\omega \hat{f}(\omega) \hat{R}(p, \omega) \exp[i\omega(t - \eta p - (h - z)P_0(p))], \quad (z < 0). \tag{13c}$$

Integrating the reflected plane waves (13c), the point-source reflection response due to the source (13a) can be written

$$\Phi_r^\omega(r, z, t)$$
$$= -\frac{i\omega}{2\pi} \hat{f}(\omega) \int_0^\infty dp\, \frac{p}{P_0(p)} \hat{R}(p, \omega) \int_0^{2\pi} d\varphi \times$$
$$\times \exp[i\omega(t - \eta p - (h - z)P_0(p))], \quad (z < 0). \tag{14a}$$

The reflection response $\Phi_r^\omega(r, z, t)$ given above is clearly seen as a composition of time-harmonic plane waves Φ_R^ω.

The Weyl-type formula (14a) for Φ_r^ω can be changed into its respective Sommerfeld-type representation by performing the φ-integration. This provides the expression

$$\Phi_r^\omega(r, z, t)$$
$$= -i\omega \hat{f}(\omega) \int_0^\infty dp\, \frac{p}{P_0(p)} \hat{R}(p, \omega) J_0(rp\omega) \exp[i\omega(t-(h-z)P_0(p))]$$
$$(r > 0;\ z < 0), \tag{14b}$$

which is well established in the Reflectivity Method (Fuchs, 1968). It no longer reveals the plane-wave components so clearly as the Weyl representation (14a).

2.4. TRANSIENT POINT-SOURCE SEISMOGRAM

According to Equation (10c), the transient point-source (1) can be considered as simulated by the (complex) incident transient plane waves

$$\Phi_I(p, x, y, z, t) = f'(t) * \Delta(t - \eta p - (z+h)P_0(p)), \quad (z+h \geq 0). \tag{15}$$

With these propagate jointly the (complex) reflected companion plane waves in the upper half-space

$$\begin{aligned}\Phi_R(p, z, y, z, t) \\ = f'(t) * R(p, t) * \Delta(t - \eta p - (h-z)P_0(p)), \quad (z < 0),\end{aligned} \tag{16}$$

where

$$R(p, t) = \mathrm{Re}\left\{\frac{1}{\pi}\int_0^\infty \hat{R}(p, \omega)\, e^{i\omega t}\, d\omega\right\}. \tag{17}$$

Hence, the (real) transient reflection response $\phi_r(r, z, t)$ for the source $f(t - R/c_0)/R$ can be written as

$$\phi_r(r, z, t) = f(t) * \mathrm{Re}\{\Psi_r(\infty, r, z, t)\}, \tag{18a}$$

where

$$\begin{aligned}\Psi_r(p^*, r, z, t) \\ = \frac{d}{dt}\left\{-\frac{1}{2\pi}\int_0^{p^*} dp\, \frac{p}{P_0(p)}\int_0^{2\pi} d\varphi\, R(p, t) * \Delta(t - \eta p - \right. \\ \left. - (h - z)P_0(p))\right\} \quad (z < 0),\end{aligned} \tag{18b}$$

from which, after the φ-integration is performed, the so-called transient Sommerfeld-type expression results

$$\begin{aligned}\Psi_r(p^*, r, z, t) \\ = \frac{d}{dt}\left\{-\frac{1}{\pi}\int_0^{p^*} dp\, \frac{p}{P_0(p)} R(p, t) * [r^2 p^2 - (t - (h-z)P_0(p))^2]^{1/2}\right\}, \\ (r > 0, z < 0).\end{aligned} \tag{18c}$$

Here the square root is as specified by condition (12).

2.5. CAUSALITY TRICK (CT) SOLUTION

As was previously shown by us (Tygel and Hubral, 1987, Ch. 7), one can simplify the reflection response (18a) by reducing the infinite integration over p to obtain

$$\phi_r(r, z, t) = f(t) * \psi_{cr}(r, z, t), \tag{19a}$$

where

$$\psi_{\mathrm{cr}}(r,z,t) = \begin{cases} 0, & (t \le 0) \\ \mathrm{Re}\{\Psi_r(p_M, r, z, t) - \overline{\Psi_r(p_M, r, z, -t)}\} & (t > 0) \end{cases} \tag{19b}$$

and

$$p_M = \max\{p_0, p_1, \ldots, p_{N+1}\}. \tag{20}$$

The response (19) results from the facts that for $p > p_M$ (a) $\mathrm{Im}\,\hat{R}(p, \omega) = 0$ and (b) $\hat{R}(p, \omega)$ has no poles on the real p axis (in fact, $|\hat{R}(p, \omega)| < 1$) for positive real values of ω (see the Appendix). These results play a key role in formulating the deconvolution problem, as will be seen below.

It is also instructive to formulate the time-harmonic version of the preceding CT result. For convenience, we recall that

$$\phi_r(r, z, t) = \mathrm{Re}\left\{\frac{1}{\pi}\int_0^\infty d\omega\, e^{i\omega t}\, \hat{\phi}_r(r, z, \omega)\right\}, \tag{21a}$$

where

$$\hat{\phi}_r(r, z, \omega) = -i\omega\int_0^\infty dp\, \frac{p}{P_0(p)}\, J_0(rp\omega)\, \hat{R}(p, \omega)\, \exp[-i\omega(h-z)P_0(p)],$$
$$(r > 0;\ z < 0). \tag{21b}$$

The CT-simplified expression for $\hat{\phi}_r$ is

$$\hat{\phi}_r(r, z, \omega) = \mathrm{Re}\{\hat{\phi}_r(r, z, \omega)\} + i\,\mathrm{Im}\{\hat{\phi}_r(r, z, \omega)\}$$
$$= H\{\mathrm{Im}\{\hat{\phi}_r(r, z, \omega)\}\} + \mathrm{Im}\{\hat{\phi}_r(r, z, \omega)\}, \tag{21c}$$

in which

$$\mathrm{Im}\{\hat{\phi}_r(r, z, \omega)\}$$
$$= \mathrm{Im}\left\{-i\omega\int_0^{p_M} dp\, \frac{p}{P_0(p)}\, \hat{R}(p, \omega)\, J_0(rp\omega)\, \exp[-i\omega(h-z)P_0(p)]\right\},$$
$$(r > 0;\ z < 0). \tag{21d}$$

In the above formulas, $H\{f\}$ denotes the Hilbert transform of f.

Note that the CT expression (21c–d) for $\hat{\phi}_r(\omega)$ involves a finite-range integral over the interval $0 \le p \le p_M$ and the evaluation of a Hilbert transform for each positive ω. This is certainly simpler to obtain than the conventional infinite-range integral over p given by the Sommerfeld representation (21b) for each positive ω.

3. Plane-Wave Decomposition

The process of extracting the plane-wave components from the point source reflected response (or from the PSS) is called a plane-wave decomposition

(PWD). This is the same, of course, as extracting the reflectivity function $R(p, t)$ (or $\hat{R}(p, \omega)$) from the data.

3.1. EXTRACTION OF TIME-HARMONIC PLANE-WAVE COMPONENTS

In the following, we use, for simplicity, Cartesian-type coordinates (x, y) and (p_1, p_2).

Let us start by writing the time-harmonic reflection response $\Phi_r^\omega(x, y, z = -h^*, t)$ due to the fundamental point-source potential $\exp[i\omega(t-R/c_0)]/R$ as a composition of time-harmonic plane-waves (PWC)

$$
\Phi_r^\omega(x, y, z = -h^*, t)
$$
$$
= -\frac{i\omega}{2\pi} \int_{-\infty}^{\infty} dp_1 \int_{-\infty}^{\infty} dp_2 \frac{p}{P_0(p)} \Phi_R^\omega(p_1, p_2, x, y, z = -h^*, t), \tag{22a}
$$

where the plane-wave components Φ_R^ω are given by

$$
\Phi_R^\omega(p_1, p_2, x, y, z = -h^*, t)
$$
$$
= \hat{R}(p, \omega) \exp[i\omega(t - p_1 x - p_2 y - (h + h^*)P_0(p))] \tag{22b}
$$

and

$$
p = \sqrt{p_1^2 + p_2^2}. \tag{22c}
$$

The above expressions were simply obtained by setting $\hat{f}(\omega) = 1$ in Equations (14a) and (13c), respectively, while transforming the cylindrical coordinates (p, φ) into Cartesian ones (p_1, p_2).

As is well known (see, e.g., Tygel and Hubral, 1987, Ch. 14), the plane-wave components Φ_R^ω at zero offset $x = y = 0$ can be obtained, for any fixed (p_1, p_2) by the so-called slant-stack operation

$$
\Phi_R^\omega(p_1, p_2, x = 0, y = 0, z = -h^*, \tau)
$$
$$
= \hat{R}(p, \omega) \exp[i\omega(\tau - (h + h^*)P_0(p))] \tag{23a}
$$
$$
= P_0(p)\left[\frac{i\omega}{2\pi} \int_{-\infty}^{\infty} dx \int_{-\infty}^{\infty} dy\, \Phi_r^\omega(x, y, z = -h^*, \tau + xp_1 + yp_2)\right]. \tag{23b}
$$

The reflection response (22a) can be viewed as a superposition of plane-waves components (22b). The slant stack or PWD (23a–b) is a process whereby each plane-wave component of the reflection response is recovered by essentially summing all traces of the shot within parallel oblique planes into a new trace at zero offset.

For fixed (p_1, p_2), the slant stack (23) is, in fact, a superposition of traces of a modified (time-delayed) reflection response of the form

$$
\tilde{\Phi}_r^\omega = \Phi_r^\omega(x, y, z = -h^*, \tau + xp_1 + yp_2). \tag{24}
$$

When integrated over the xy-plane, this modified reflection response Φ_r^ω collapses to the plane-wave component Φ_R^ω specified by (p_1, p_2), at the symmetry axis $x = y = 0$.

In order to better understand the equation pair (22)–(23), it is instructive to rewrite Equation (22a) as

$$\Phi_r^\omega(x, y, z = -h^*, t)$$
$$= -\frac{i\omega}{2\pi} \int_{-\infty}^{\infty} dp_1 \int_{-\infty}^{\infty} dp_2 \, \Lambda(p_1, p_2, z = -h^*, t - xp_1 - yp_2), \quad (25a)$$

where

$$\Lambda(p_1, p_2, z = -h^*, \tau)$$
$$= \frac{\hat{R}(p, \omega)}{P_0(p)} \exp[i\omega(\tau - (h + h^*)P_0(p))] \quad (25b)$$

$$= \frac{1}{P_0(p)} \Phi_R^\omega(p_1, p_2, x = 0, y = 0, z = -h^*, \tau). \quad (25c)$$

In other words, the PWC (25a) can be seen as a superposition of plane-wave components with a multiplicative factor $1/P_0(p)$ and time delay $\tau - p_1 x - p_2 y$ with respect to zero offset $x = y = 0$.

Recall that the reflection response $\Phi_r^\omega(x, y, z = -h^*, \tau)$, on the other hand, was time delayed by $\tau + p_1 x + p_2 y$ in the PWD or slant stack formulas (23).

In this way, it seems natural that $\Lambda(p_1, p_2, z = -h^*, \tau)$ and $\Phi_r^\omega(x, y, z = -h^*, t)$ correspond to each other in a transform pair. (This transform pair is nothing but the so-called Radon transform, which is now so much applied in tomography).

In summary, the Radon transform pair connects the reflection response $\Phi_r^\omega(x, y, z = -h^*, t)$ to the zero offset plane-wave components $\Phi_R^\omega(p_1, p_2, x = 0, y = 0, z = -h^*, \tau)$ multiplied by the factor $1/P_0(p)$. Note also the change in sign of the factor $i\omega/2\pi$ is the PWC and PWD formulas.

Introducing cylindrical coordinates in Equations (23) and performing the angle integral in the transformed formula, we obtain the simple expression

$$\Phi_R^\omega(p, x = 0, y = 0, z = -h^*, \tau)$$
$$= \hat{R}(p, \omega) \exp[i\omega(\tau - (h+h^*)P_0(p))]$$
$$= P_0(p)(i\omega) \int_0^{\infty} dr \, r \, J_0(rp\omega) \, \Phi_r^\omega(r, z = -h^*, \tau), \quad (r > 0). \quad (26)$$

This expression can be considered as the inversion formula of Equation (14b). Although formula (26) is operationally simpler than formulas (23), it no longer reveals so easily the PWD process. Equations (24) and (26) may be referred to as being of the Weyl and Sommerfeld-type, respectively.

3.2. EXTRACTION OF TRANSIENT PLANE-WAVE COMPONENTS

Multiplying both parts of Equation (26) by $1/\pi$ and integrating over positive ω, we can derive the transient version of that formula, namely

$$R(p, \tau) * \Delta(\tau - (h+h^*)P_0(p))$$

$$= P_0(p) \frac{d}{d\tau}\left\{\frac{1}{\tau}\int_0^\infty dr\, r\, \Phi_r(r, z = -h^*, \tau) * \mathrm{Re}[r^2p^2 - \tau^2]^{-1/2}\right\}. \qquad (27)$$

It is instructive to comment a little on the above important formula. Taking the real parts of both sides and calling

$$\phi_r(r, z = -h^*, \tau) = \mathrm{Re}\{\Phi_r(r, z = -h^*, \tau)\}, \qquad (28a)$$

we can write Equation (27) as

$$R(p, \tau) * \mathrm{Re}\left\{\frac{1}{P_0(p)}\,\omega(\tau - (h+h^*)P_0(p))\right\}$$

$$= \frac{d}{d\tau}\left\{\frac{1}{\pi}\int_0^\infty dr\, r\, \phi_r(r, z = -h^*, \tau) * \mathrm{Re}[r^2p^2 - \tau^2]^{-1/2}\right\}. \qquad (28b)$$

Now, if $0 \leq p \leq p_0$, then $P_0(p) = |P_0(p)|$ is real. Hence,

$$\mathrm{Re}\left\{\frac{1}{P_0(p)}\Delta(\tau - (h+h^*)P_0(p))\right\} = \frac{1}{P_0(p)}\delta(\tau - (h+h^*)P_0(p)). \qquad (29a)$$

On the other hand, when $p > p_0$, $P_0(p) = -i|P_0(p)|$ is purely imaginary, so that

$$\mathrm{Re}\left\{\frac{1}{P_0(p)}\Delta(\tau - (h+h^*)P_0(p))\right\}$$

$$= \frac{-\tau}{\pi[\tau^2 + (h+h^*)^2|P_0(p)|^2]|P_0(p)|}. \qquad (29b)$$

It follows that
(a) If $0 \leq p \leq p_0$, then

$$R(p, \tau) = P_0(p)\frac{d}{d\tau}\left\{\frac{1}{\pi}\int_0^\infty dr\, r\, \phi_r(r, z = -h^*, \tau + (h+h^*)P_0(p)) *\right.$$

$$\left. * \mathrm{Re}[r^2p^2 - \tau^2]^{-1/2}\right\} \qquad (29c)$$

and
(b) If $p > p_0$, then

$$R(p, \tau) * \left\{\frac{-\tau}{\pi[\tau^2 + (h+h^*)^2|P_0(p)|^2]}\right\}.$$

$$= |P_0(p)|\frac{d}{d\tau}\left\{\frac{1}{\pi}\int_0^\infty dr\, r\, \phi_r(r, z = -h^*, \tau) * \mathrm{Re}[r^2p^2 - \tau^2]^{-1/2}\right\}. \qquad (29d)$$

In case (a) (subcritical) $R(p, \tau)$ is obtained directly from the data. In case (b) one needs a deconvolution to recover $R(p, \tau)$. We now turn to the problem of recovering the unknown source pulse from the reflection seismogram.

4. Deconvolution Procedure

Let us consider in the following the reasonable condition

$$c_0 < c_{N+1} \text{ (or } p_{N+1} < p_0) \tag{30}$$

that is, the velocity in the lower half-space is higher than the velocity in the upper layer.

Let us also introduce the following terminology: we call the range $\{p_{N+1} < p < p_0\}$ the *amplitude spectrum stack range* and $\{p > p_M = \max(p_0, \ldots, p_{N+1})\}$ the *phase spectrum stack range*.

As data, we consider the point-source seismogram (PSS)

$$\phi_S(r, z = -h^*, t) = f(t - R^*/c_0)/R^* + \phi_r(r, z = -h^*, t), \tag{31a}$$

where

$$R^* = \sqrt{r^2 + (h - h^*)^2}. \tag{31b}$$

We assume the arbitrary pulse or wavelet $f(t)$ as well as the reflectivity function $R(p, t)$ of the stratified model, provided that condition (30) holds. From the theory explained in Sections 2.1 and 2.3, Equations (2a–f) in Section 1.1 should become clear. Also recall from Section 1.1 the transformed data function $\hat{G}(p, \omega)$ (Equations (3a–b)) which is obtained from the PSS $\phi_S(r, z = -h^*, t)$ by means of an FT of parameter (frequency) ω followed by a Hankel transform of 0-order of ray parameter p. A procedure similar to the one that led to Equation (25) can be used to readily establish the fundamental Equation (4).

Assume now the properties (5a–b) of the time-harmonic reflectivity function $\hat{R}(p, \omega)$ (which are proved in the Appendix) and select any ray parameter $\bar{p}$ in the phase spectrum stack range, i.e. $\bar{p} > p_M = \max(p_0, \ldots, p_{N+1})$. We can write

$$\hat{f}(\omega) M(\bar{p}, \omega) = \hat{G}_S(\bar{p}, \omega) \tag{32a}$$

where

$$M(\bar{p}, \omega) = -\frac{-i}{\omega P_0(\bar{p})} [e^{-i\omega|h-h^*|P_0(\bar{p})} + \hat{R}(\bar{p}, \omega) e^{-i\omega(h-h^*)P_0(\bar{p})}]$$

$$= \frac{1}{\omega|P_0(\bar{p})|} [e^{\omega|h-h^*||P_0(\bar{p})|} + \hat{R}(\bar{p}, \omega) e^{\omega(h+h^*)|P_0(\bar{p})|}]. \tag{32b}$$

As can be readily verified, the fact that $\hat{R}(\bar{p}, \omega)$ is real and $|\hat{R}(\bar{p}, \omega)| < 1$ (Equation (5a)) implies that

$$M(\bar{p}, \omega) > 0 \quad \text{(for all } \omega > 0). \tag{32c}$$

Taking principal arguments $(-\pi < \arg \alpha \leq \pi)$ on both sides of Equation (4), it follows from the elementary property of the argument of the product that

$$\arg\{\hat{f}(\omega)\} = \arg\{\hat{G}_S(\bar{p}, \omega)\} \tag{32d}$$

since we have, of course,

$$\arg M(\bar{p}, \omega) = 0. \tag{32e}$$

This establishes Equation (6c).

We now proceed to prove Equations (6d) and (6e) and, for this matter, select $\tilde{\tilde{p}}$ in the range of the amplitude spectrum stack, namely $p_{N+1} < \tilde{\tilde{p}} < p_0$. From Equation (4), we can write

$$\hat{f}(\omega) R_0(\tilde{\tilde{p}}, \omega) e^{-i\omega(h+h^*) P_0(\tilde{\tilde{p}})}$$
$$= \hat{G}_S(\tilde{\tilde{P}}_0) i\omega P_0(\tilde{\tilde{p}}) - \hat{f}(\omega) e^{-i\omega|h-h^*| P_0(\tilde{\tilde{p}})}. \tag{33a}$$

Note that $P_0(p)$ is real, so that

$$\left|e^{-i\omega(h+h^*) P_0(\tilde{\tilde{p}})}\right| = \left|e^{-i\omega|h-h^*| P_0(\tilde{\tilde{p}})}\right| = 1. \tag{33b}$$

Moreover, from property (5b) of the reflectivity function $\hat{R}(p, \omega)$, we have

$$\left|\hat{R}(\tilde{\tilde{p}}, \omega)\right| = 1. \tag{33c}$$

Taking absolute values for both sides of Equation (33a) and taking into consideration Equations (33b–c), we have the relationship

$$\left|\hat{f}(\omega)\right| = \left|\hat{G}_S(\tilde{\tilde{p}}, \omega) i\omega P_0(\tilde{\tilde{p}}) - \hat{f}(\omega) e^{-i\omega|h-h^*| P_0(\tilde{\tilde{p}})}\right|. \tag{34}$$

Squaring both sides and using the elementary formula of complex numbers

$$|z_1 - z_2|^2 = |z_1|^2 + |z_2|^2 - 2|z_1||z_2| \cos[\arg z_1 - \arg z_2], \tag{35}$$

we find, after some manipulation, Equations (6d–e).

The determination of $\hat{f}(\omega)$ and $\hat{R}(p, \omega)$ from which $f(t)$ and $R(p, t)$ are obtained is shown by Equations (6g–i).

By now, the reader should, of course, have a feeling why we have called the ranges $p > p_M = \max(p_0, \ldots, p_{N+1})$ and $p_{N+1} < p < p_0$ the amplitude and phase stack spectrum ranges, respectively. Using $\bar{p}$ and $\tilde{\tilde{p}}$ in these ranges, we were able to recover the phase and amplitude of $\hat{f}(\omega)$, respectively.

5. Conclusions

In this chapter, we have given a procedure by which one can recover both the wavelet and the reflectivity functions from a point-source seismogram of a stratified acoustic medium. We assumed the condition that the velocity in the lower half-space is higher than the velocity in the upper half-space (a condition that is almost always met is practice) and the source point to be located in the upper half-space.

Generalizations can be made for arbitrary locations of the point-source within the medium.

The concepts of amplitude and phase stack spectrum ranges were introduced. They shed more light on the role of inhomogeneous plane-wave components in the representation of the point-source.

The use of supercritical plane-wave components provides the possibility of recovering the arbitrary wavelet and reflectivity because of the important properties of the time-harmonic reflectivity functions in this range.

We are confident that more results can be obtained by looking at the plane-wave components in the supercritical region. This range has been rejected in the past on such grounds such as its effect is negligible in the seismogram and other, often unjustified, reasons. The supercritical range is the only one that easily offers strong properties on the reflectivity function and this is certainly of value, as this work has tried to indicate. After first submitting this paper we learned that Ziolkowski *et al.* (1987) considered extracting a *dynamite* (i.e., minimum-phase) wavelet with the help of the critical reflection theorem from real data. The theory proposed in this chapter shows that, in principle, it should also be possible to extract an *arbitrary* mixed-delay wavelet from real data.

Appendix: Fundamental Properties of Time-Harmonic Reflectivity

In this appendix we prove properties (5a–b) of the time-harmonic reflectivity function $\hat{R}(p, \omega)$ that were used in the text. For convenience, we rewrite them here recalling that the condition $c_0 < c_{N+1}$ (or $p_{N+1} < p_0$) is assumed. Namely, we wish to prove that for all $\omega > 0$ the time-harmonic reflectivity $\hat{R}(p, \omega)$ satisfies

(a)   If $p > p_M = \max(p_0, \ldots, p_{N+1})$ (i.e., for p belonging to the phase stack spectrum range), $\hat{R}(p, \omega)$ is real and $|\hat{R}(p, \omega)| < 1$.

and

(b)   If $p_{N+1} < p < p_0$ (i.e., for p belonging to the amplitude stack spectrum range), then $|\hat{R}(p, \omega)| = 1$.

Statements (a) and (b) are proved by induction using the 'bottom to top' recursion scheme to obtain $\hat{R}(p, \omega)$. This recursion reads (see, e.g., Tygel and Hubral, 1987, Ch. 6)

$$\hat{R}_N = r_N \qquad\qquad\qquad\qquad\qquad\qquad\text{(A.1a)}$$

and

$$\hat{R}_j = \frac{r_j + z_{j+1}\hat{R}_{j+1}}{1 + r_j z_{j+1}\hat{R}_{j+1}} \quad (j = N-1, \ldots, 1, 0). \tag{A.1b}$$

The functions $\hat{R}_j = \hat{R}_j(p, \omega)$ are the time-harmonic reflectivity of the stack of layers below the interface $z = H_j$ assuming a half-space above this interface with the acoustic parameters (c_j, ρ_j) of layer j. It is then clear that

$$\hat{R}(p, \omega) = \hat{R}_0(p, \omega). \tag{A.1c}$$

Also,

$$r_j = \frac{\rho_{j+1}P_j(p) - \rho_j P_{j+1}(p)}{\rho_{j+1}P_j(p) + \rho_j P_{j+1}(p)} \quad (j=0, 1, \ldots, N) \tag{A.2a}$$

is the reflection coefficient of interface $z = H_j$ from above and

$$z_{j+1} = \begin{cases} \exp[-2i\omega(H_{j+1} - H_j)P_{j+1}(p)], & (j=0, 1, \ldots, N-1) \\ 1, & (j = N) \end{cases} \tag{A.2b}$$

relates to the (complex) vertical transient time for traversing layer $j+1$ ($j=0, 1, \ldots, N-1$). The value $z_{N+1} = 1$ is just a useful notation.

In connection with the above formulas, we prove the following fundamental result that leads to statements (a) and (b).

THEOREM. *For all $j=N, N-1, \ldots, 1, 0$ and for all $\omega > 0$, we have*

(I) *If $p \geq \max(p_j, p_{N+1})$ and $\hat{R}_j(p, \omega)$ is finite, then*

$$\overline{\hat{R}}_j(p, \omega) = \hat{R}_j(p, \omega). \tag{A.3a}$$

(Here the overbar denotes complex conjugate.)

(II) *If $p_{N+1} \leq p \leq p_j$, then*

$$\overline{\hat{R}}_j(p, \omega) = 1/\hat{R}_j(p, \omega). \tag{A.3b}$$

(III) *If $p > \max(p_j, \ldots, p_{N+1})$ then*

$$-1 < \hat{R}_j(p, \omega) < 1. \tag{A.3c}$$

Statement (I) tells that if $p \geq \max(p_j, p_{N+1})$ and $\hat{R}_j(p, \omega)$ is not singular (p is not a pole), then $\hat{R}_j(p, \omega)$ is real. Statement (II) indicates that if $p_{N+1} \leq p \leq p_j$, then $|\hat{R}_j(p, \omega)| = 1$. (This is sometimes referred to as the total reflection theorem (Fokkema and Ziolkowski, 1985.) Statement (III) indicates that $\hat{R}_j(p, \omega)$ is real and $|\hat{R}_j(p, \omega)| < 1$ if $p > \max(p_j, \ldots, p_{N+1})$.

As indicated above, statements (I), (II) and (III) are proved by induction on $j=N, \ldots, 1, 0$. As can be easily verified, these assertions hold for $\hat{R}_N = r_N$.

Assume now that they hold for some R_{j+1} $(0 \leq j < N)$. Our task is, then, to show their validity for $\hat{R}_j$.

Suppose $p \geq \max(p_j, p_{N+1})$ and $R_j(p, \omega)$ finite. We then consider the two possibilities $p \geq p_{j+1}$ and $p < p_{j+1}$.

If $p \geq p_{j+1}$, we have clearly $\bar{r}_j = r_j$, $\bar{z}_{j+1} = z_{j+1}$. If $\hat{R}_{j+1}(p, \omega)$ is infinite, then $\hat{R}_j = 1/r_j$ is real. If $\hat{R}_{j+1}(p, \omega)$ is finite then, by induction hypothesis, $\bar{\hat{R}}_{j+1} = \hat{R}_{j+1}$.

Taking conjugates of both sides of Equation (A.1b), we have

$$\bar{\hat{R}}_j = \frac{\bar{r}_j + \bar{z}_{j+1} \bar{\hat{R}}_{j+1}}{1 + \bar{r}_j \bar{z}_{j+1} \bar{\hat{R}}_{j+1}} \tag{A.4}$$

from which substitution in the above yields $\hat{R}_j = \bar{\hat{R}}_j$ as claimed.

If $p < p_{j+1}$, then clearly $\bar{r}_j = 1/r_j$, $\bar{z}_{j+1} = 1/z_{j+1}$ and, by induction, $\bar{\hat{R}}_{j+1} = 1/\hat{R}_{j+1}$. Substituting in Equation (A.4) yields, once more, $\hat{R}_j = \bar{\hat{R}}_j$. This proves statement (I).

To prove (II), we consider $p_{N+1} \leq p \leq p_j$ and the cases $p \geq p_{j+1}$ and $p < p_{j+1}$.

If $p \geq p_{j+1}$, then $\bar{r}_j = 1/r_j$, $\bar{z}_{j+1} = z_{j+1}$. If $\hat{R}_{j+1}(p, \omega)$ is infinite, then $\hat{R}_j = 1/r_j$ from which $\bar{\hat{R}}_j = 1/\hat{R}_j = r_j$ as claimed. If $\hat{R}_{j+1}(p, \omega)$ is finite, then, by induction, $\bar{\hat{R}}_{j+1} = 1/\hat{R}_{j+1}$. Substitution of the above in Equation (A.4) produces $\bar{\hat{R}}_j = 1/\hat{R}_j$.

If $p < p_{j+1}$, then $\bar{r}_j = r_j$, $\bar{z}_{j+1} = 1/z_{j+1}$ and, by induction, $\bar{\hat{R}}_{j+1} = 1/\hat{R}_{j+1}$. Substituting on (A.4) yields, once more, $\bar{\hat{R}}_j = 1/R_j$, as desired. Thus, statement (II) is proved.

We finally turn to prove (III). For this matter, we write

$$1 \pm \hat{R}_j = 1 \pm \frac{r_j + z_{j+1}\hat{R}_{j+1}}{1 + r_j z_{j+1}\hat{R}_{j+1}} = (1 \pm r_j)\left[\frac{1 \pm z_{j+1}\hat{R}_{j+1}}{1 + r_j z_{j+1}\hat{R}_{j+1}}\right]. \tag{A.5}$$

We now observe that for $p > \max(p_j, \ldots, p_{N+1})$ we have

$$|r_j| < 1 \tag{A.6a}$$

and

$$1 \pm z_{j+1} R_{j+1} \geq 1 - z_{j+1}|\hat{R}_{j+1}| > 0. \tag{A.6b}$$

Conditions (A.6) follow easily from the definition of the reflection coefficient r_j, the quantity z_{j+1} and the induction hypothesis on $\hat{R}_{j+1}$. Note that $|z_{j+1}| = z_{j+1} < 1$ if $p > \max(p_j, \ldots, p_{N+1})$.

Hence, for $p > \max(p_j, \ldots, p_{N+1})$,

$$1 \pm \hat{R}_j > 0, \tag{A.7}$$

which is equivalent to (III).

Acknowledgements

We thank Prof. Lúcio Tunes dos Santos for useful discussions and carefully reading the manuscript. This work has been financially supported by the Federal German Ministry of Research and Technology (BMFT).

References

Bube, K. P. and Burridge, R., 1983, The one-dimensional inverse problem of reflection seismology, *SIAM Rev.* **25**, 497–559.

Brekhovskikh, L. M., 1980, *Waves in Layered Media*, Academic Press, New York.

Fokkema, J. and Ziolkowski, A., 1987, The critical reflection theorem, *Geophysics* **52**, 965–972.

Fuchs, K., 1986, The reflection of spherical waves from transition zones with arbitrary depth-dependent elastic moduli and density, *J. Phys. Earth* **16**, 27–41.

Kelamis, P. G., Chiburis, E. F., and Friedmann, V., 1987, Post-critical wavelet estimation and deconvolution, in *Technical Programme of the EAEG Meeting at Belgrade, Yugoslavia, 9–12 June 1987.*

Müller, G., 1971, Direct inversion of seismic observations, *J. Geophys.* **37**, 225–235.

Poritzky, H., 1951, Extension of Weyl's integral for harmonic spherical waves to arbitrary wave shapes, *Comm. Pure Appl. Math.* **4**, 33–42.

Robinson, E. A., 1954, Predictive decomposition of time series with applications to seismic exploration, PhD Thesis, MIT, Cambridge, Mass. Also in *Geophysics* **32**, 418–484 (1967).

Sonnevend, G. Y., 1987, Sequential and stable methods for the solution of mass recovery problems (estimation of the spectrum and of the impedance function). Paper presented at the 5th International Seminar on Model Optimization in Exploration Geophysics, Berlin.

Symes, W. W., 1983, Impedance profile inversion via the finite transport equation, *J. Math. Anal. Appl.*, **94**, 435–453.

Tygel, M. and Hubral, P., 1987, *Transient Waves in Layered Media*, Elsevier, Amsterdam.

Ursin, B. and Berteussen, K. A., 1986, Comparison of some inverse methods for wave propagation in layered media, Proceedings of the IEEE, **74**, 389–400.

Weyl, H., 1919, Ausbreitung elektromagnetischer Wellen über einen ebenen Leiter, *Ann. Phys.* **60**, 481–500.

Yagle, A. E. and Levy, B. C., 1984, Application of the Schur algorithm to the inverse problem for a layered acoustic medium, *J. Acoust. Soc. Am.* **76**, 301–308.

Ziolkowski, A. M., Fokkema, J. T., Baeten, G. J. M., and Ras, P. A. W., 1987, Extraction of the dynamite wavelet on real data using the critical reflection theorem, in *Technical Programme of the EAEG Meeting at Belgrade, Yugoslavia, 9–12 June 1987.*

M. Tygel
Department of Applied Mathematics,
IMECC/UNICAMP,
131000 Campinas,
Sao Paulo,
Brazil

P. Hubral and F. Wenzel
Geophysical Institute,
University of Karlsruhe,
Kaiserstrasse 12,
7500 Karlsruhe,
West Germany

ANTON ZIOLKOWSKI, JACOB T. FOKKEMA,
KLAAS-JAN KOSTER, ARJEN CONFURIUS
and RUUD VAN BOOM

Inversion of Common Mid Point Seismic Data

Abstract

Inversion of seismic reflection data is normally done by iterative forward modelling. There are three problems with this approach. First, the choice of initial model and the number of parameters required to specify it are usually unconstrained. Secondly, because of the lack of constraints, it becomes extremely difficult to find the optimum model. The optimum model is usually found by minimizing the error between the wavefield calculated from the model and the measured wavefield. The minimum reached is usually a local, rather than a global, minimum. Thirdly, the data are used only to calculate the error.

In our view, the choice of initial model should be inherently constrained by the data. We have developed a scheme that works in either the time domain or the frequency domain, for determining the acoustic parameters of the layered earth by directly inverting the CMP data, assuming that a CMP gather is equivalent to a shot gather over a horizontally-layered acoustic earth. The scheme is recursive, and is essentially the inverse of the forward reflectivity method. The recovered earth model is constructed from the top layer downwards, by stripping off the layers one by one. All interbed peg-leg multiples are included in the scheme. The result is an equivalent plane layered earth model for each CMP gather.

To apply the scheme in practice it is crucial to have accurate measurements of the source wavefield. If the source is not a point, the wavefield must be collapsed to that of a point source before transform of the data to plane wave components in the τ-p domain. The source wavelet and the influence of the free surface must be removed before applying the layer-stripping algorithm. The algorithm is unstable in the presence of noise. Stabilization is achieved by limiting the magnitude of the recovered reflection coefficients, and by limiting the number of peg-leg multiples that are removed with each layer.

We show that the scheme works on synthetic data but, in the presence of noise, the recovered impedance profile is in error at low frequencies. This low frequency information can be obtained from the travel-time information using normal velocity analysis and an empirical velocity-density relation. Since this trend information is essentially independent of the layer-stripping results, it is probably better to use the frequency domain method, in which the trend information does not influence the layer-stripping scheme.

Paul L. Stoffa (ed.), Tau-p: A Plane Wave Approach to the Analysis of Seismic Data, 141–176.
© 1989 *by Kluwer Academic Publishers.*

We have also applied our scheme to tank data and to a shot record from a data set shot in the North Sea. In both cases, the inversion works well and is stable. Addition of the trend information to recover the impedance must be done with great care. The recovery of the reflection coefficient series is robust and independent of the trend information. Comparison of the inverted North Sea data with a well-log is quite good. Errors in the inversion are caused both by lack of precision in the specification of the acquisition parameters, especially of the source wavefield, and in the removal of the free surface, which is done by predictive deconvolution. There are also errors in the log.

Introduction

GENERAL CONSIDERATIONS

In its most general sense, the 'inversion of seismic data' means the following: Given the seismograms and all required information about the source and receiver configuration, calculate the elastic parameters of the earth. The corresponding forward problem may be stated as follows: Given the elastic parameters of the earth and a configuration of source and receivers, calculate the seismograms that would be generated.

The object of inversion is thus very easy to understand. Unfortunately it is much harder to achieve the objective than to state it. This is because, in practice, the inverse problem is not simply the inverse of the forward problem. There are three basic limitations of seismic data that make inversion difficult. First, it is impossible to measure the whole wavefield. In practice, the wavefield is measured only at a number of places at or near the earth's surface (and sometimes in a well). Secondly, the wave generated by the seismic source is bandlimited. Thirdly, the data are always noisy. Any sensible scheme of seismic inversion must take these limitations into account.

Consider a computer program that solves the forward problem: it calculates the seismic waves in a modelled three-dimensional elastic earth in response to an impulsive point source placed at some point in the earth. The earth could be modelled as a large number of elements each with its own elastic parameters. The program could solve the problem by calculating the progress of the waves through the earth model as a sequence of time steps. At any time step, each point in the earth would either be at rest — before the arrival of the fastest wave from the source — or would be moving in response to the seismic wave. Now consider the inverse problem: Given the motion of every point in the earth for all times, and the position of the source, calculate the elastic parameters of the earth. It is conceivable that this problem could be solved, and it is possible that there is a unique set of parameters that would give the described motions. But we will not worry about this, because *this* inverse problem never arises: it is impossible to measure the motion of every point in the earth.

Suppose it would be possible to measure the motion of every point on the *surface* of the earth, with infinite bandwidth, with no errors, and with no noise. Would it then be possible to determine the elastic parameters of every point *within* the earth? As far as we know, this problem has not been solved, and it is open to conjecture whether there is a unique answer to this problem. In fact, we are able to measure the motion of the earth only over a finite bandwidth, with errors, with noise, and *only in a few discrete places*. There is thus a fundamental problem of uniqueness. That is, there is more than one set of elastic parameters for the earth that is consistent with the measurements. Probably there is an infinite number of sets of such parameters. Perhaps some of these sets fit the data but are physically and geologically implausible: these can be excluded. There is thus a limited range of earth models that can be accepted. In our opinion the way in which this acceptable range of earth models is defined is the central problem in designing an inversion scheme. (For a discussion of the uniqueness problem for an earth model that is constrained to be spherically symmetric, see Aki and Richards (1980), pp. 699–706, and Backus and Gilbert (1970).)

The second limitation is bandwidth. The bandwidth of the source sets a limit to the resolution available. That is, the size of the elements within the earth that can be described by the measurements is inversely proportional to the bandwidth. This problem is now very well understood (Berkhout, 1984).

The final limitation is noise. Not only do the measurements have errors (because the behaviour of the transducers is imperfectly known and not absolutely reproducible), there is always added noise. The noise level in seismology is always far greater than the smallest measurable seismic signal and is a much more important consideration than the linearity, repeatability, or dynamic range of the instruments. The effect of noise is to increase the magnitude of the errors in the measurements. It therefore reduces the *precision* with which the elastic parameters in any element can be specified. Any inversion scheme must recognize that noise limits the precision with which the earth can be described. (Noise also introduces numerical problems because of the finite signal bandwidth, but this is a separate issue.)

Thus, because we have a limited number of measurements, a finite bandwidth, and noise, we have a problem of uniqueness, a limit to the obtainable resolution, and a lack of precision. In seismic *migration* the seismic data are imaged to form a model of the earth, taking these limitations into account. However, migration can only be performed if the *velocities* are known. The objective of inversion is to *determine* the velocities and densities of every point of the earth.

INVERSION BY MODEL-FITTING

A number of so-called inversion schemes are really iterative forward-modelling schemes in which the wavefield generated by an initial model is computed and compared with the real data. If the fit of the synthetic data is an acceptable match to the true data, the final model has been obtained. If the match between the synthetic and real data is not optimum, in some sense, the model is updated, or perturbed in some way, to make the fit better. When further iterative updates give no significant improvement, the optimum model has been reached. An outline of this general scheme is shown in Figure 1.

There are three basic problems with this scheme. First, and most importantly, there is infinite freedom in the choice of the initial model: this is unconstrained. Secondly, the updating of the model in order to reduce the error is a nonlinear problem. It is customary (see, for example, Berkhout, 1987) to linearize the problem around the current estimate and then to adjust the model parameters to minimize the error. There is no guarantee that the minimum reached is the global minimum. It is likely to be a local minimum. Treitel (1987) explained in an oral presentation that entirely different optimum models can be derived with this scheme, which satisfy the goodness of fit criteria equally well (or equally badly), and the optimum model that is reached depends on the initial model. In other words, local minima are reached. Variations in the outcome can also be obtained by changing the way the goodness of fit (or the error) is defined. Thirdly, the data are used only to estimate the error.

In order to make the scheme more likely to converge on the global minimum, it is clearly important to start with a good initial model. This was recognized by Tarantola (1984). A good global initial model can be obtained from the measured data using the travel time information to estimate the velocities, and using an empirical velocity-density relation (Gardner et al., 1974) to obtain the densities. However, this is essentially low frequency information, and does not contain detail. The detail is supposed to come out of the iterative forward modelling scheme, and it will only come out if the initial model is sufficiently close to the

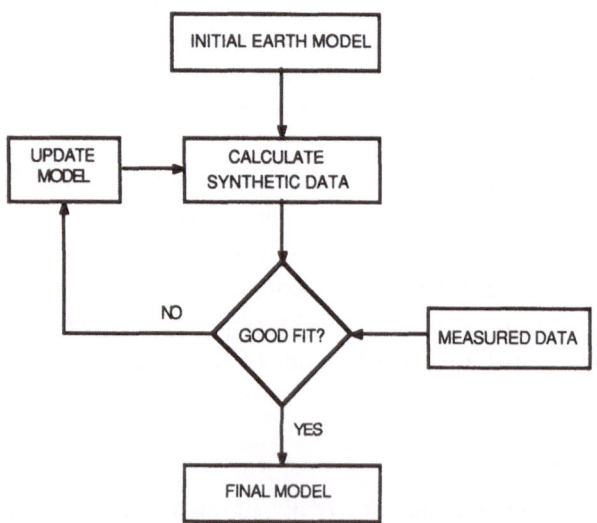

Fig. 1.   Diagram showing scheme for inversion by iterative forward modelling.

truth. If the initial model is not close to the true earth, the details that emerge may, to some extent, be spurious. Another important consideration in the definition of the initial model is the number of parameters required to specify it. This should be intimately related to the bandwidth, the signal-to-noise-ratio of the data, and the number of recorded data. In other words, it is data-dependent.

In our view, the freedom we have to choose the initial model for inversion by iterative forward modelling should be inherently constrained by the data. This inherent constraint should be built into the derivation of the model from the data. That is, the earth model should be derived from the data *directly*. In order to force this direct relationship between the derived (or inverted) earth model and the data, it is essential to construct an analytic forward model of the seismic wavefield. If bandlimited synthetic data calculated using this model can be inverted, in the presence of noise, to recover the original model (with errors of course), then the scheme can be applied to real data to recover an inverted earth model. This recovered model could then either be refined in an iterative forward modelling scheme, or accepted as the final model.

The basic limitation of our approach is that we must severely constrain our earth model in order to obtain an analytical description of the wavefield. We are looking for ways to relax the constraints we impose. What we present here is our initial approach to the problem.

The Forward Problem

INTRODUCTION: THE MODEL

In our approach we recognize the tremendous success that common mid point stacking (Mayne, 1962) has had in the seismic industry, and propose to use an earth model that exploits the features of the common mid point (CMP) gather that have led to this success. The reason that CMP stacking works so well is that a CMP gather looks like a shot gather over a plane horizontally-layered earth (Robinson and Treitel, 1980, p. 10; Diebold and Stoffa, 1981). That is, the arrival times of the reflections in a CMP gather are more or less the same as those that would be obtained from a shot gather over an equivalent plane-layered earth structure. Since each CMP gather is slightly different from its neighbours, we propose a different equivalent plane layered earth model for each CMP gather.

In this approach, therefore, we argue that the earth can be regarded approximately as one-dimensional in the CMP domain. With this approximation we have a sufficient constraint to find a unique solution to the inversion problem. Inversion, in this approach, amounts to the determination of the layer parameters of the equivalent horizontally layered earth for each CMP gather. We consider here only the acoustic case, but the scheme we propose may in principle be extended to the full elastic case for isotropic layered media (Du Cloux, 1987).

The theory in this section is from Fokkema and Ziolkowski (1987) and rests on the reflectivity method (Fuchs and Muller, 1971). Our initial model is an acoustic plane layered earth bounded by upper and lower half spaces. We consider individual plane wave components of the incident wave and study the structure of the response. We then modify the model, replacing the upper half space by a layer with a free surface, and introduce a point source in this upper layer. The incident field of the point source may be expressed as a sum of plane wave components, and the wavefield in the upper layer may be expressed as the incident field plus a sum of plane wave components scattered by the layered half space. The free surface imposes an interference pattern on the scattered response, introducing peaks and notches into the spectrum. The positions of the notches depend on the depth of the source and receiver below the free surface, and the velocity in the upper layer, while the positions of the peaks depend on the thickness of the layer.

PLANE WAVE RESPONSE

We consider a stack of N plane parallel acoustic layers, as shown in Figure 2,

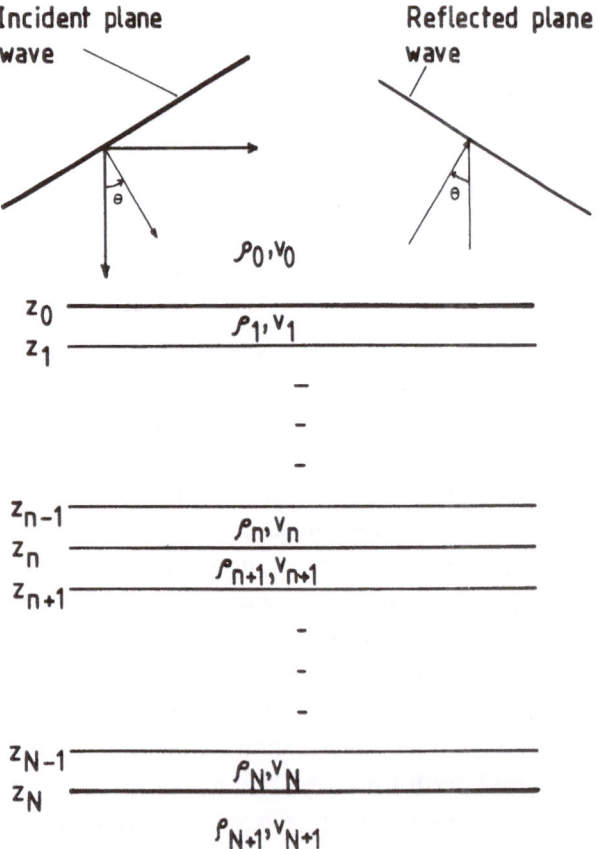

Fig. 2. A plane horizontally-layered acoustic earth model sandwiched between two half spaces. A plane wave is incident from the upper half space at an angle θ to the vertical.

bounded at the bottom by a homogeneous acoustic halfspace of density ρ_{N+1} and velocity v_{N+1}, and at the top by a homogeneous acoustic half space of density ρ_0 and velocity v_0, where v_0 is less than v_{N+1}. A plane pressure wave is incident from the upper half space. The normal to the wavefront is parallel to the x–z plane and is incident at an angle θ to the normal to the layers, which is chosen to be the z axis.

The wave propagation is analyzed in the space-frequency domain (x, z, ω) and the complex time factor $\exp(-i\omega t)$ is omitted in the equations. In our subsequent discretized inversion scheme, the angular frequency ω is defined as $\omega = 2\pi m \, \Delta f$, where Δf is the frequency increment.

The incident field is the plane pressure wave

$$P^{INC}(x, z, \omega) = A_0^+(\omega) \exp\{i\omega(p_0 x + q_0 z)\}, \tag{1}$$

where $A_0^+(\omega)$ is the spectrum of the pressure wave function,

$$p_0 = \sin \theta / v_0 \tag{2}$$

is the horizontal slowness, or ray parameter, and

$$q_0 = (1/v_0^2 - p_0^2)^{1/2} \tag{3}$$

is the vertical slowness in the upper half space. The reflection response of the stratified medium is unknown, but it will be a wave returning at an angle θ to the normal, propagating upwards:

$$P^R(x, z, \omega) = A_0^-(\omega) \exp\{i\omega(p_0 x - q_0 z)\}. \tag{4}$$

The total pressure field in the upper half space is the sum of the incident and reflected fields

$$P^0(x, z, \omega) = \exp(i\omega p_0 x)[A_0^+(\omega) \exp(i\omega q_0 z) + A_0^-(\omega) \exp(-i\omega q_0 z)]. \tag{5}$$

The normal component of particle acceleration is related to the pressure gradient by Newton's second law of motion. In the time domain this may be expressed as

$$\frac{d^2 u_z(x, z, t)}{dt^2} = -\frac{1}{\rho} \frac{\partial p(x, z, t)}{\partial r}, \tag{6}$$

where $u_z(x, z, t)$ is the normal component of particle displacement. For a plane wave the total derivative with respect to time d/dt is equal to the partial derivative $\partial/\partial t$, and Equation (6) may be transformed to the frequency domain to yield

$$U_z(x, z, \omega) = \frac{1}{\rho\omega^2} \frac{\partial P(x, z, \omega)}{\partial z}. \tag{7}$$

Thus, in the upper half space, layer 0, the normal component of displacement is

$$U_z^0(x, z, \omega) = \frac{iq_0}{\rho_0 \omega} \exp(i\omega p_0 x)[A_0^+(\omega) \exp(i\omega q_0 z)$$
$$- A_0^-(\omega) \exp(-i\omega q_0 z)]. \tag{8}$$

Similarly, the pressure and normal displacement in the nth and $n + 1$st layers are

$$P^n(x, z, \omega) = \exp(i\omega p_n x)[A_n^+(\omega) \exp(i\omega q_n z)$$
$$+ A_n^-(\omega) \exp(-i\omega q_n z)], \tag{9}$$

$$U_z^n(x, z, \omega) = \frac{iq_n}{\rho_n \omega} \exp(i\omega p_n x)[A_n^+(\omega) \exp(i\omega q_n z)$$
$$- A_n^-(\omega) \exp(-i\omega q_n z)], \tag{10}$$

for all values of x and for $z_{n-1} \le z \le z_n$, and

$$P^{n+1}(x, z, \omega)$$
$$= \exp(i\omega p_{n+1} x)[A_{n+1}^+(\omega) \exp(i\omega q_{n+1} z) + A_{n+1}^-(\omega) \exp(-i\omega q_{n+1} z)], \tag{11}$$

$$U_z^{n+1}(x, z, \omega) = \frac{iq_{n+1}}{\rho_{n+1} \omega} \exp(i\omega p_{n+1} x) \times$$
$$\times [A_{n+1}^+(\omega) \exp(i\omega q_{n+1} z) - A_{n+1}^-(\omega) \exp(-i\omega q_{n+1} z)], \tag{12}$$

for all values of x and $z_n \le z \le z_{n+1}$.

From the continuity of pressure at the boundaries for all values of x, it follows that

$$p_0 = p_1 = \ldots = p_n = p_{n+1} = \ldots = p_{N+1}, \tag{13}$$

which is Snell's law and can be expressed as

$$p_n = \frac{\sin \theta}{v_0} = p, \quad \text{for } n = 0, 1, 2, \ldots, N + 1. \tag{14}$$

Since

$$q_n = (1/v_n^2 - p_n^2)^{1/2}, \tag{15}$$

q_n will become imaginary if $p > 1/v_n$.

At each boundary pressure and displacement are continuous. That is

$$\lim_{z \uparrow z_n} P^n(x, z, \omega) = \lim_{z \downarrow z_{n+1}} P^{n+1}(x, z, \omega), \tag{16}$$

$$\lim_{z \uparrow z_n} U^n(x, z, \omega) = \lim_{z \downarrow z_{n+1}} U^{n+1}(x, z, \omega) \tag{17}$$

We now define a global reflection coefficient $R_n(\omega)$ for the nth layer, as follows

$$A_n^-(\omega) = R_n(\omega) A_n^+(\omega) \exp(2i\omega q_n z_n). \tag{18}$$

Divide Equation (17) by Equation (16), after substitution from expressions (9), (10), (11), and (12), using Equation (18) and a similar expression for the $n + 1$st layer to obtain the following recursion formula for the global reflection coefficient

$$R_n(\omega) = \frac{\Gamma_n + R_{n+1}(\omega) \exp(i\omega \tau_{n+1})}{1 + \Gamma_n R_{n+1}(\omega) \exp(i\omega \tau_{n+1})}, \tag{19}$$

in which

$$\tau_{n+1} = 2q_{n+1}(z_{n+1} - z_n) \tag{20}$$

is the vertical two-way travel time in layer n, and

$$\Gamma_n = \frac{(q_n/\rho_n) - (q_{n+1}/\rho_{n+1})}{(q_n/\rho_n) + (q_{n+1}/\rho_{n+1})} \tag{21}$$

is the local reflection coefficient. (It can be seen that when $p = 0$, $q_n = 1/v_n$, $q_{n+1} = 1/v_{n+1}$, and Equation (21) is the well known expression for the reflection coefficient at normal incidence.) Using definition (18) for the upper half space, Equation (5) may be written as

$$P^0(x, z, \omega) = A_0^+(\omega) \exp(i\omega px)[\exp(i\omega q_0 z) + R_0(\omega) \exp\{i\omega q_0(2z_0 - z)\}]. \tag{22}$$

POINT SOURCE RESPONSE AND INFLUENCE OF THE FREE SURFACE

In order to relate the plane wave results to the seismic reflection method, we now use the earth model shown in Figure 3, in which the upper half space of Figure 2 has been replaced by a layer with a free surface at $z = 0$. This upper layer has the same velocity and density as the upper half space of Figure 2. We introduce a monopole source in this layer at a depth h_s on the z axis.

The pressure in this uppermost layer consists of an incident field P^{INC} and a

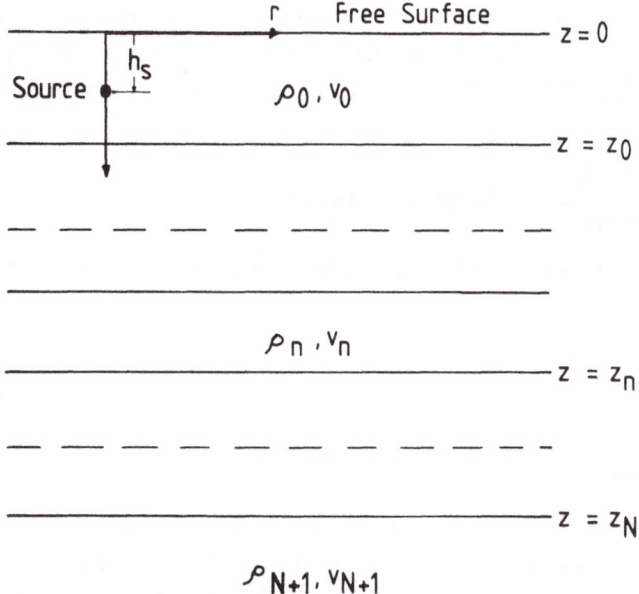

Fig. 3.   Introduction of a point source and a free surface. The earth model is the same as in Figure 2, except that the upper half space has been replaced by an acoustic layer with a free surface, and a point source has been introduced into this layer.

scattered field P^S:

$$P_0(p_x, p_y, z, \omega) = P^{INC}(p_x, p_y, z, \omega) + P^S(p_x, p_y, z, \omega), \tag{23}$$

in which we introduce the third dimension y. The incidence field satisfies the inhomogeneous wave equation

$$\frac{\partial^2 P^{INC}}{\partial z^2} + \omega^2 q_0{}^2 P^{INC} = -S(\omega)\ \delta(z - h_s), \tag{24}$$

in which

$$q_0 = \left(\frac{1}{v_0{}^2} - p_x{}^2 - p_y{}^2\right)^{1/2} \tag{25}$$

from cylindrical symmetry, and $S(\omega)$ is the source spectrum in this domain. The well known solution of Equation (24) is

$$P^{INC} = \frac{iS(\omega)}{2\omega q_0}\ \exp[i\omega(p_x x + p_y y + q_0|z - h_s|)], \quad \text{for } 0 \le z \le z_0. \tag{26}$$

The factor $\exp\{i\omega(p_x x + p_y y)\}$ is common in both the incident and scattered fields and will be omitted in further equations for simplicity.

The scattered field consists of upgoing and downgoing waves, as in other layers:

$$P^S = A_0^+(\omega)\ \exp(i\omega q_0 z) + A_0^-(\omega)\ \exp[i\omega(2q_0 z_0 - q_0 z)] \tag{27}$$

for $0 \le z \le z_0$.

We now find the global reflection coefficient $R_0(\omega)$ at the boundary $z = z_0$. As before, this is the ratio of the upgoing to the downgoing wave, but now recognizing that the downgoing wave consists of the incident wave and the scattered wave

$$A_0^-(\omega) = R_0(\omega)\left[\frac{iS(\omega)}{2\omega q_0}\ \exp(-i\omega q_0 h_s) + A_0^+(\omega)\right]. \tag{28}$$

At the free surface $z = 0$, the boundary condition is that the pressure is zero:

$$P_0 = 0. \tag{29}$$

Therefore from Equations (23), (26), and (27), it follows that

$$A_0^+(\omega) = -1\left[\frac{iS(\omega)}{2\omega q_0}\ \exp(i\omega q_0 h_s) + A_0^-(\omega)\ \exp(2i\omega q_0 z_0)\right] \tag{30}$$

in which the well known -1 reflection coefficient at the free surface is clearly recognizable. Equations (28) and (30) may be solved for A_0+ and A_0-, thus enabling the scattered field in Equation (27) to be determined. Combining this result with Equation (26) yields the complete expression for the total field in layer

0 at depth level z, as follows:

$$P_0(p, z, \omega) = S(\omega)G(p, z, \omega) \tag{31}$$

in which

$$p = (p_x^2 + p_y^2)^{1/2} \tag{32}$$

and the Green's function $G(p, z, \omega)$ is given by

$$G(p, z, \omega) = \frac{i}{2\omega q_0} [G^{INC}(p, z, \omega) + G^S(p, z, \omega)], \tag{33}$$

where

$$G^{INC}(p, z, \omega) = \exp[i\omega q_0|z - h_s|] - \exp[i\omega q_0(z + h_s)] \tag{34}$$

and

$$G^S(p, z, \omega) = \frac{R_0(\omega) \exp[i\omega q_0(2z_0 - z - h_s)]}{1 + R_0(\omega) \exp(2i\omega q_0 z_0)} [1 - \exp(2i\omega q_0 h_s)] \times$$
$$\times [1 - \exp(2i\omega q_0 z)]. \tag{35}$$

The Green's function consists of an incident field and a scattered field. The incident field contains two terms, described by Equation (34): the first term is the direct wave from the source, and the second term is the reflection of this wave in the free surface (the 'ghost'). The scattered field is described by Equation (35) and contains five factors: $R_0(\omega)$ is of course the global reflection coefficient at $z = z_0$ and contains all the information about the layered halfspace; the denominator describes the multiple behaviour in the first layer; the factor $[1 - \exp(2i\omega q_0 h_s)]$ is the ghost operator at the source; the factor $[1 - \exp(2i\omega q_0 z)]$ is the ghost operator at the receiver; and $\exp[i\omega q_0(2z_0 - z - h_s)]$ is simply a phase factor.

We may also find an expression for the vertical component of the particle velocity $-i\omega U_z$ using the relation

$$-i\omega U_z(p, z, \omega) = \frac{1}{i\omega\rho_0} \frac{\partial P_0(p, z, \omega)}{\partial z}. \tag{36}$$

The interesting case for exploration geophysics is for $z = 0$, corresponding to the situation where geophones are placed on the surface, and we find

$$\frac{\omega^2 \rho_0 U_z(p, 0, \omega)}{S(\omega)} = \exp(i\omega q_0 h_s) +$$
$$+ \frac{R_0(\omega) \exp[i\omega q_0(2z_0 - h_s)]}{1 + R_0(\omega) \exp(i\omega q_0 z_0)} [1 - \exp(2i\omega q_0 h_s)]. \tag{37}$$

The direct wave and its free surface reflection combine to give a single term $\exp(i\omega q_0 h_s)$ for the incident field. In the scattered field the multiples and ghost operator are still there, but the receiver ghost vanishes as the geophones are at the surface.

The Inverse Problem

INTRODUCTION

The inverse problem now consists of determining the layer parameters, given the reflection response of each plane wave component. This can be done recursively by stripping off layers, starting at the top, an approach which is numerically unstable. The source of the instability is inaccuracies in the calculation and subtraction of the multiples in the presence of noise. The recognition of this problem has led us to develop a stabilized recursive layer-stripping scheme.

We first present the layer-stripping scheme and then discuss a processing scheme to recover the reflection response of each plane wave component from the seismic reflection data.

THE INVERSE REFLECTIVITY METHOD

The forward calculation of the plane wave response of a horizontally layered earth uses Equations (19), (20) and (21). The calculation starts at the bottom of the sequence with the observation that there are no upgoing waves in the lower half space. The calculation is thus initialised with $R_N(\omega) = \Gamma_N$. It proceeds recursively by adding layers $N-1$, $N-2$, ..., etc. to construct the global reflection coefficient $R_0(\omega)$ for the whole sequence.

The inversion scheme is in principle exactly the reverse of this forward calculation. It begins with $R_0(\omega)$ and recursively works downwards by stripping off layers to obtain $R_1(\omega)$, $R_2(\omega)$, etc. A crucial constraint on the scheme is that it must work on the pre-critically reflected plane wave components. The reason is that the reflection response is *causal* if the angle of incidence of the incident plane wave is less than the critical angle in the fastest layer. The property of causality is explicitly exploited in our scheme. For critically and post-critically incident plane waves the reflection response is non-causal: the head waves arrive before the incident wave.

To apply the inversion scheme in practice, we first rewrite Equation (19) as

$$R_{n+1}(\omega) \exp(i\omega\tau_{n+1}) = \frac{\Gamma_n - R_n(\omega)}{\Gamma_n R_n(\omega) - 1} = -\Gamma_n + X_n(\omega), \tag{38}$$

where

$$X_n(\omega) = \frac{(1 - \Gamma_n^2)R_n(\omega)}{1 - \Gamma_n R_n(\omega)}. \tag{39}$$

We now discretize the layers using the Goupillaud principle

$$\tau_n = \Delta t \quad \text{for all } n, \tag{40}$$

and remember that for our discrete data $\omega = 2\pi m \, \Delta f$, where Δf is the frequency increment. Equations (38) and (39) may now be written as

$$R_{n+1}(m \, \Delta f) \exp(2\pi i m \, \Delta f \, \Delta t) = -\Gamma_n + X_n(m \, \Delta f) \tag{41}$$

in which

$$X_n(m \, \Delta f) = \frac{(1 - \Gamma_n^2)R_n(m \, \Delta f)}{1 - \Gamma_n R_n(m \, \Delta f)}. \tag{42}$$

In principle, the inverse problem is now solved. However, since we have not taken into account the noise or the finite bandwidth, this scheme is very unstable. Small errors in $R_n(m \, \Delta f)$ will rapidly accumulate because of the recursion and the denominator in Equation (42) may even become zero. The problem may be understood and made more tractable in the time domain.

STABILIZATION IN THE TIME DOMAIN

Transforming Equation (42) to the time domain yields

$$x_n(M \, \Delta t) - \Gamma_n \sum_{m=0}^{M} r_n((M - m) \, \Delta t)x_n(m \, \Delta t)$$
$$= (1 - \Gamma_n^2)r_n(M \, \Delta t), \quad \text{for } M = 0, 1, 2, \ldots, M_{max} \tag{43}$$

The first arrival, for $M = 0$, is the primary reflection. Thus

$$r_n(0) = \Gamma_n \tag{44}$$

is the local reflection coefficient. Equation (43) can now be written as

$$x_n(M \, \Delta t) = r_n(M \, \Delta t) + \beta \sum_{m=0}^{M-1} r_n((M - m) \, \Delta t)x_n(m \, \Delta t), \tag{45}$$

with

$$x_n(0) = r_n(0), \tag{46}$$

and

$$\beta = \Gamma_n/(1 - \Gamma n^2), \tag{47}$$

and we see that $x_n(M \, \Delta t)$ is obtained recursively. We may transform the relation (41) to the time domain and, using (44), we obtain

$$r_{n+1}(m \, \Delta t) = x_n((m + 1) \, \Delta t). \tag{48}$$

In this scheme, we have two means by which we stabilize the inversion. First, when the local reflection coefficient r_n is small. β is also small and, hence, the correction on $r_n(m\,\Delta t)$ is small. If this small correction is ignored, then $r_{n+1}(m\,\Delta t)$ is no more than a shifted version of $r_n(m\,\Delta t)$. The physical meaning of this is that a small reflection coefficient generates very small multiple reflections that are indistinguishable from the noise. We therefore choose β as a threshold for the detection in the presence of noise of significant reflectors and their peg-leg multiples.

Secondly, the summation term in Equation (45) removes the peg-leg multiples. The peg-leg multiple train is a time sequence that has coefficients that decrease in magnitude with increasing time. After a certain length of time, this multiple train will be buried in the noise and the calculation of the multiple train, and its removal will be essentially calculations based on noise, rather than on signal. Therefore, it is sensible to limit the time window over which the multiples are calculated and subtracted. To achieve this the full summation from 0 to $M-1$ in Equation (45) is replaced by a shorter summation, from 0 to L, where $L < M-1$:

$$x_n(M\,\Delta t) = r_n(M\,\Delta t) + \beta \sum_{m=0}^{L} r_n((M-m)\,\Delta t)\,x_n(m\,\Delta t), \qquad (49)$$

in which L is another threshold variable. Thus the two variables β and L may be chosen to stabilize the layer-stripping scheme.

We may summarise the scheme as follows:

TWT (two-way time) $= 0$: $r_0(t) \rightarrow \Gamma_0 = r_0(0)$ [Eq. (44)]

[Eq. (49)]

$x_0(t)$

[Eq. (48)]

TWT $= \Delta t$: $r_1(t) \rightarrow \Gamma_1 = r_1(0)$ [Eq. (44)]

[Eq. (49)]

$x_1(t)$

[Eq. (48)]

TWT $= 2\Delta t$: $r_2(t) \rightarrow \Gamma_2 = r_2(0)$ [Eq. (44)]

etc.

STABILIZATION IN THE FREQUENCY DOMAIN

The principles that are required to stabilize the inversion in the time domain may also be applied in the frequency domain. We see first that Equation (44) may be written as

$$\Gamma_n = r_n(0) = \sum_{m=-M_{max}}^{M_{max}-1} R_n n(m\,\Delta f).$$ (50)

Equation (49) was derived from Equation (42). The problem was with the denominator which could become very small. We can write Equation (42) as

$$X_n(m\,\Delta f) = (1 - \Gamma_n^2) R_n(m\,\Delta f)[1 - \Gamma_n R_n(m\,\Delta f)]^{-1}$$

$$= (1 - \Gamma_n^2) R_n(m\,\Delta f)[1 + \Gamma_n R_n(m\,\Delta f) - \Gamma_n^2 R_n^2(m\,\Delta f) + \cdots]$$ (51)

in which the multiple train is the infinite series in the square brackets. The coefficients of the series rapidly decrease since, for pre-critical reflection, both $|\Gamma_n|$ and $|R_n(m\,\Delta f)|$ are less than one. (That $|\Gamma_n|$ is less than one can be seen from Equation (21); the proof that $|R_n(m\,\Delta f)|$ is less than one is a corollary of the critical reflection theorem (Fokkema and Ziolkowski, 1987).) Therefore, in the presence of noise, only the first few terms in the series are significant. And, clearly, if $|\Gamma_n|$ is very small, Equation (51) reduces to

$$X_n(m\,\Delta f) = (1 - \Gamma_n^2) R_n(m\,\Delta f).$$ (52)

The Fourier transform of Equation (48) is

$$R_{n+1}(m\,\Delta f) = X_n(m\,\Delta f) \exp(-2\pi\iota\,\Delta f\,\Delta t).$$ (53)

In the frequency domain the scheme has exactly the same structure as in the time domain:

TWT (two-way time) = 0: $R_0(m\Delta f) \rightarrow r_0(0) = \Gamma_n$ [Eq. (50)]

        [Eq. (51)]

        $X_0(m\Delta f)$

        [Eq. (53)]

TWT = Δt: $R_1(m\Delta f) \rightarrow r_1(0) = \Gamma_1$ [Eq. (50)]

        [Eq. (51)]

        $X_1(m\Delta f)$

        [Eq. (53)]

TWT = $2\Delta t$: $R_2(m\,\Delta f) \rightarrow r_2(0) = \Gamma_2$ [Eq. (50)]

        etc.

The stabilized frequency domain scheme is faster than the time domain version. It also has the advantage that the *number* of peg-leg multiples to be removed is specified, rather than the *time duration* of this multiple train. This has a more 'scientific' feel to it.

Data Processing Scheme

The data processing scheme is as follows:

(1) Convert recorded data to measured parameter
(2) Sort data into CMP gathers
(3) Remove surface waves (land data)
(4) Collapse source array to a point
(5) Decompose CMP gathers into plane waves
(6) (Calibrate seismic data with well data)
(7) Deconvolve source wavelet
(8) Remove multiples generated by top layer
(9) Apply layer stripping inversion scheme to obtain reflection coefficients
(10) Determine acoustic parameters
(11) Repeat for all CMP gathers

In this scheme there are a number of steps that require explanation. Not all the steps have been completely worked out and, at this time, approximations to the ideal solutions must be made.

Step 1, converting recorded data to the measured parameter, is not normally done. In the equations for the inversion process absolute values of pressure or particle velocity are required. Thus the recorded data, normally volts on tape, should be converted to pressure or particle velocity, taking into account the sensitivity of the transducers and the response of the recording system.

Step 2, sorting into CMP gathers, is essential for seismic data in areas where the dip is not negligible. In areas of neglible dip the raw shot gathers can be used. The essential point of this step is that the data must look like shot gathers over a plane horizontally layered earth, and thus the earth may be regarded as having variations in only the vertical direction.

Step 3: remove surface waves. Surface waves are not included in our theory. Therefore any surface waves (ground roll) should be removed. Every effort should be made in acquisition not to generate surface waves. However, if they are recorded, the spatial sampling interval should be small enough to prevent aliasing and permit them to be removed by velocity filtering. It is possible that where the data are spatially aliased, the plane wave decomposition (step 5) will move the surface waves to very large ray parameters corresponding to the evanescent region for upgoing and downgoing waves.

Step 4, collapsing the source array to a point, is not a conventional data processing step, and it needs some explaining. In the analysis of the forward problem, we decomposed a point source into an infinite number of plane wave components and defined the Green's function in the (p, z, ω) domain, where p, the horizontal slowness or ray parameter, was defined by Equation (32). The analysis exploits the cylindrical symmetry about the source point by choosing the z-axis to go through the source. In order to decompose the wavefield into plane waves, it is essential that the source be a point.

A monopole point source, such as we have assumed in our model, can be approximated in practice by using a source whose dimensions are small compared with the wavelengths of acoustic radiation that are generated. In the marine case this would be true of, for example, an air gun or a water gun. In the land case it would be true of a single charge of dynamite. A single vibrator on the earth's surface could be regarded as a dipole source. In nearly all marine seismic data, arrays of sources are used. These arrays are not small compared with the wavelengths of radiation they generate and do not have a wavefield that has the required cylindrical symmetry. Something must be done to correct this in processing. The same would be true of most vibroseis data in which arrays of vibrators are normally used as the seismic source.

For two-dimensional data, in which the geophone or hydrophone cable is in the x-direction, the source should have small dimensions in the x-direction. This is not normally the case. If it *were* the case, the source could be regarded as a spatial delta-function $\delta(x)$ with a spatial wavenumber spectrum that would be constant for all spatial wavenumbers k_x. If the source is an array, it will have some directivity. In the farfield of the array, that is, at distances large compared with the array dimensions, the Fraunhofer approximation to the directivity pattern can be made. The directivity pattern can be described as a spatial Fourier transform of the array aperture function at any frequency. The source spectrum can thus be described in frequency and wavenumber as $S(\omega, k_x)$. The collapse of the source array to a point can be regarded as removing the k_x dependence of the source wavefield from the data. Because the data can be regarded as shot gathers over horizontal layers, the downgoing and upgoing wavenumbers are the same. Therefore, in the far field of the source, the spatial deconvolution of the seismic reflection data may be accomplished by a complex division per frequency in the wavenumber domain. The result will then be an approximation to data that would have been obtained with a point source of spectrum $S(\omega)$. It goes without saying that we assume the source wavefield is known from independent measurements. In the case of a marine seismic source array these measurements can be made as proposed by Ziolkowski *et al.* (1982), and by Parkes *et al.* (1984).

Step 5, decomposition of the wavefield into plane waves may now be done using the Hankel transform

$$P_0(p, z, \omega) = \int_0^\infty \bar{P}(r, z, \omega) J_0(\omega p r) r \, \mathrm{d}r, \tag{54}$$

in which the pressure data measured at depth level z, offset r and frequency ω

are decomposed into the plane wave components. The plane wave data now have
the form described by Equation (31), in which the Green's function is described
by Equations (33), (34) and (35).

Step 6, calibration of the seismic data with well data, is only necessary if it is not
possible to recover the measured parameter from the recorded data (step 1). A
synthetic seismogram is calculated in the τ-p domain from the known wavelet
and well logs according to the forward modelling scheme described above and in
Fokkema and Ziolkowski (1987). The real seismogram is then scaled to the
synthetic such that, for example, the energy in the two seismograms is the same
within the signal bandwidth.

Step 7, deconvolution of the source wavelet, simply means dividing $P_0(p, z, \omega)$
by $S(\omega)$. Care must be taken to avoid division by zero; so this step may be taken
as follows:

$$G(p, z, \omega) = \frac{P_0(p, z, \omega).S^*(\omega)}{|S(\omega)|^2 + C}, \tag{55}$$

in which the asterisk denotes complex conjugate, and C is a small real constant.

Step 8: remove multiples generated by top layer. This is nontrivial. After step 7,
we have the Green's function which consists of the incident field and the scattered
field. Very often we can neglect the incident field in our data because our
hydrophone groups and geophone groups are very insensitive to energy arriving
end-on; they are designed to have maximum sensitivity to energy arriving broad-
side. Thus our Green's function can be regarded as being given by Equation (35),
from which we need to extract $R_0(\omega)$.

What we should really like to do is to remove all the effects of the free surface
using some wave theoretical approach. This would then leave us with $R_0(\omega)$.
Unfortunately we do not know how to do this yet, and must make approximations. The two ghost factors in the numerator essentially act as bandpass filters:
we cannot do much about their influence on the amplitude spectrum, but we can
remove their influence on the phase spectrum by division in the frequency
domain. The multiples are generated by the denominator. This denominator can
be removed in principle if the thickness and velocity of the top layer are known.
In practice this is known to be very unstable unless the thickness and velocity are
known with great precision. We resort to the conventional approach for removal
of first-layer multiples: predictive deconvolution.

Step 9. In principle we now have $R_0(\omega)$ and may apply our layer stripping
inversion scheme to obtain the reflection coefficients.

Step 10, determination of the acoustic parameters. Step 9 may be applied to
each of the plane wave components in the data. The result is essentially the
reflection coefficient series for each plane wave component and contains primary
reflections only. It should be mentioned that not all the plane wave components
can be used. Since the causality of the response is explicitly exploited in the
inversion scheme, only the precritically reflected plane wave components can be

used; that is, for ray parameters not exceeding $1/v_{max}$, where v_{max} is the largest velocity in the layered sequence. In this step the primary interval velocities and densities are obtained.

Velocity analysis in the τ-p domain is now well established (Schultz, 1976; Stoffa et al., 1981, 1982; Stoffa, 1985). The difference between the data after step 9 and normal seismic data transformed to the τ-p domain, is that the data after step 9 contain only primary reflections. The source wavelet has been removed by deconvolution and all multiple reflections have been removed by the layer-stripping process.

Step 11. The above processing sequence must be repeated for every CMP gather. The result will be the equivalent layer parameters of the primary reflectors as a function of two-way travel time. The data may then be imaged by migration using the well-known exploding reflector model.

Addition of Trend Information

It is well known that pre-critically reflected seismic data lack low frequency energy below about 5 Hz, and very often even below 10 Hz. At these low frequencies the wavelengths of the seismic waves are large compared with the thickness of any layer in the earth; or, conversely, all layers within the earth appear thin at such low frequencies, and such thin layers have a reflection response that decreases linearly with decreasing frequency (Widess, 1973; Koefoed and De Voogd, 1980; Ziolkowski and Fokkema, 1986). It follows that information about the interior of the earth at such low frequencies cannot be recovered from the data using the inversion scheme described above.

This low frequency information concerns the well-known general increase of velocities and densities with depth. The low frequency velocity information is obtained using the normal methods of velocity analysis that correlate events on seismograms recorded at different offsets. Such travel-time information usually yields a rough layered earth model with about 10 layers. The densities of these layers can be estimated approximately using, for example, the formula of Gardner et al. (1974). A synthetic seismogram computed using such a model will contain both low and high frequency energy because the layer boundaries are sharp. Only the low frequencies can be trusted because the velocity analysis is not reliable at high frequencies. Therefore the high frequencies should be filtered out of this synthetic seismogram. Any high frequency information about the velocities, above say 5 Hz, should come out of the analysis in step 10.

Of course if a well is available, both high and low frequency information can be obtained at the well site. The high frequency part is liable to change rapidly with distance from the well; the low frequency part is likely to change more slowly. Very often information about velocities and densities is not obtainable from the shallowest layers in the well, because of the well casing. In this shallow part velocities estimated from the refracted seismic data can be used.

Results

SYNTHETIC DATA

Figure 4 shows a density log and a velocity log from a North Sea well. The plane wave synthetic reflection response for normal incidence was calculated using the forward modelling scheme described above and this was convolved with the minimum-phase bandlimited wavelet shown in Figure 5. Noise was then added to the seismogram. The result was the synthetic equivalent of the plane wave response after Step 6 in the data processing sequence described above, with the difference that no free surface effect has to be removed from the synthetic. The known wavelet was deconvolved as described by Equation (55), and the reflection coefficient series was recovered using the stabilized time domain recursion described above. The results are shown in Figure 6 for 10 and 30 % rms noise.

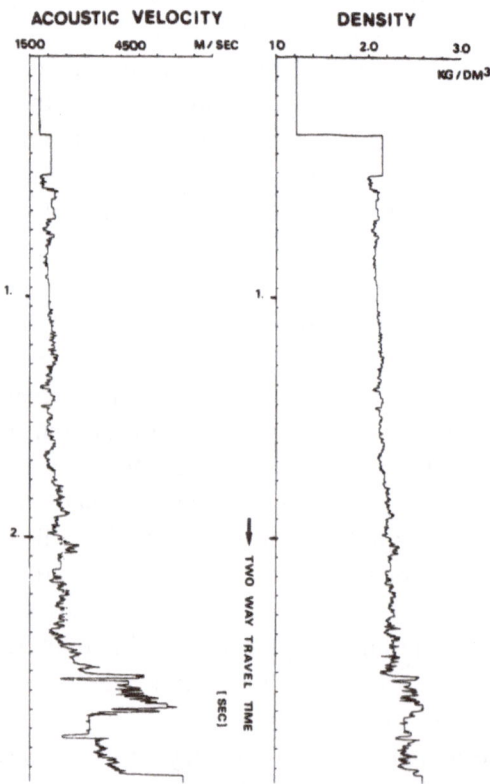

Fig. 4. Density and velocity logs from a North Sea well. The upper half of the density log has been obtained from the upper half of the velocity log using a modification of the formula of Gardner *et al.* (1974). The first few thick layers in the velocity log were interpreted from refracted arrivals over the well.

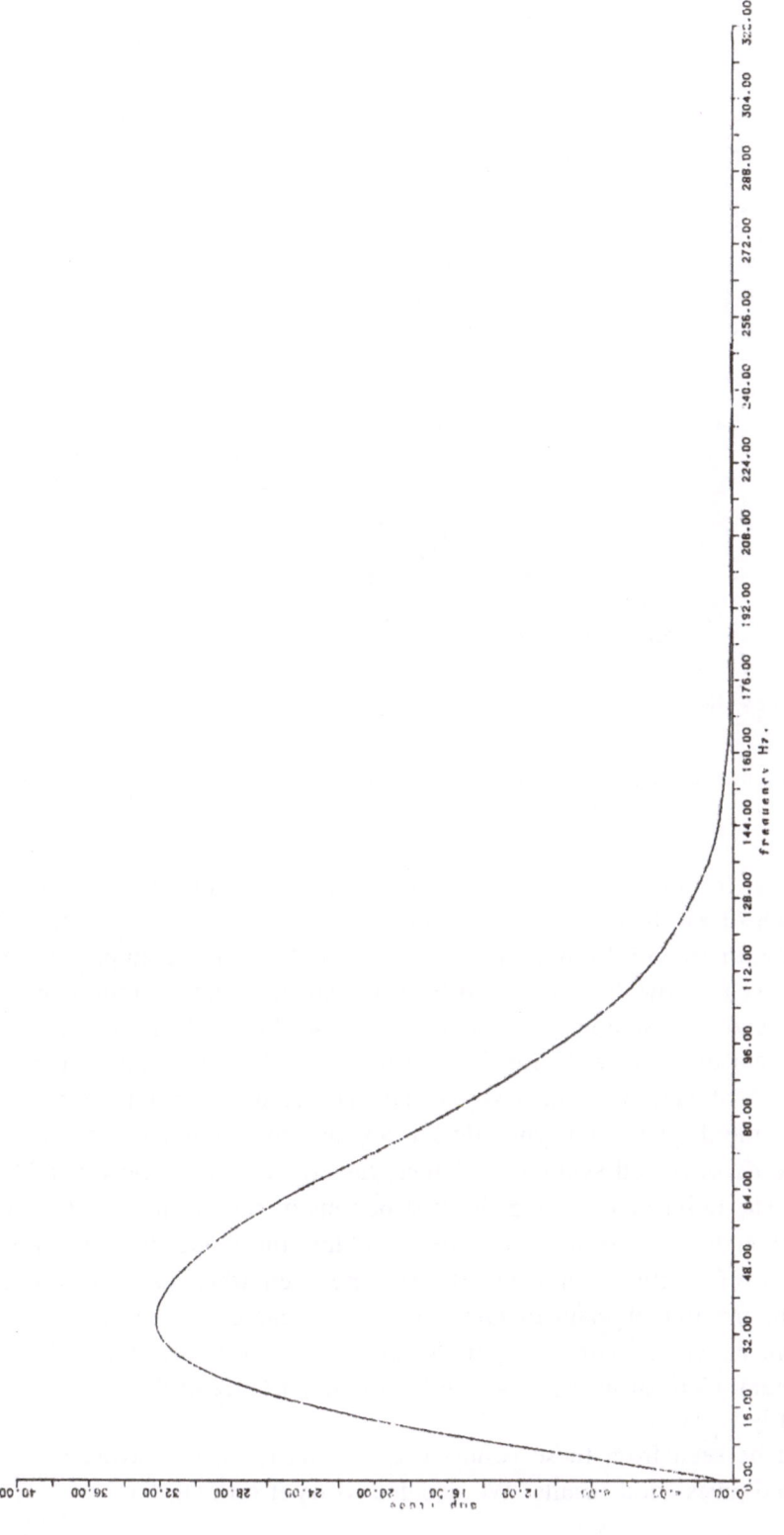

Fig. 5.   Bandlimited seismic wavelet used to compute synthetic seismograms.

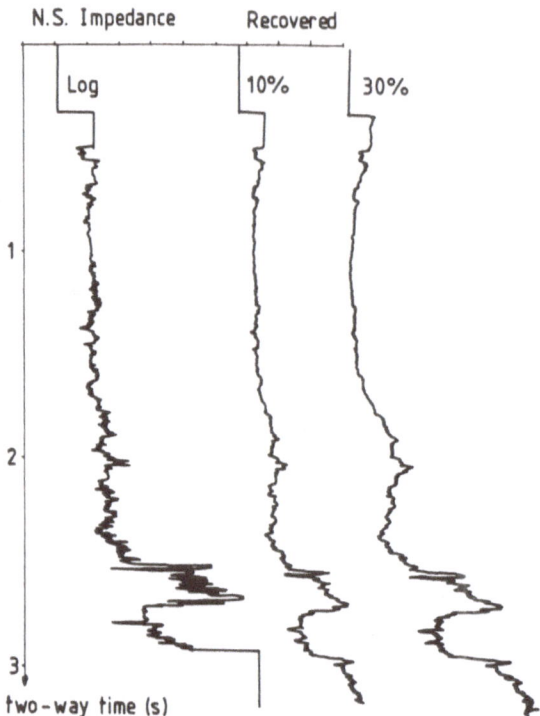

Fig. 6. Recovered acoustic impedance profiles from noisy synthetic seismograms versus well log impedance profile, for 10 and 30 % noise.

It is clear from Figure 6 that the algorithm is stable in the presence of noise and that much of the higher frequency information is recovered correctly. The low frequency, or trend information, is not recovered. From the amplitude spectrum of the wavelet shown in Figure 5 it is clear that there is very little energy below 8 Hz. Trend information was obtained from stacking velocities estimated from seismic data over the well site. The calculated reflection response (without the wavelet of course) was low-pass filtered. The resultant trend seismogram was added to the deconvolved synthetic seismogram, after removing the first 600 ms from the deconvolved synthetic. That is, the trend seismogram was added after Step 8. The point of removing the first 600 ms of the synthetic data is to show that, although the scheme is recursive, building the model from the top down, the trend information can guide the scheme even when data are absent. The results of the time-domain inversion of these trend-corrected synthetic seismograms are shown in Figure 7. Figure 8 shows the result of inverting only the trend seismogram. Comparing Figures 6 and 7, the importance of the trend information can clearly be seen.

It can be seen from these results that the kernel of the inversion scheme is stable and gives good results. We decided to try it on real data.

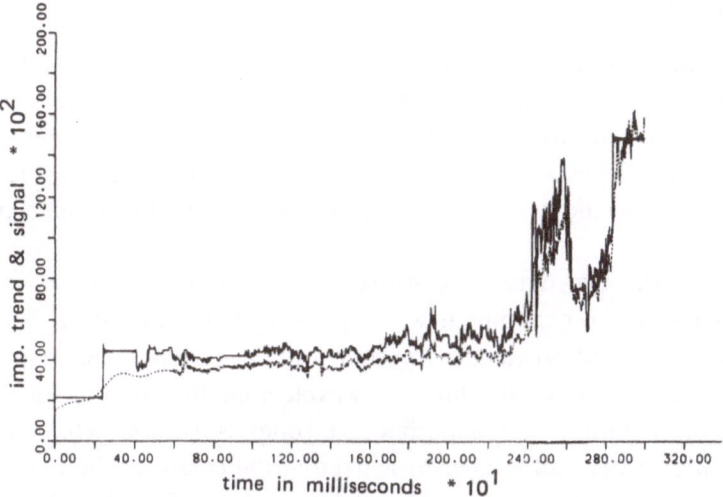

Fig. 7. Recovered acoustic impedance profiles from noisy synthetic seismograms after first, the removal of the first 600 ms of the seismograms and, second, addition of the trend seismogram.

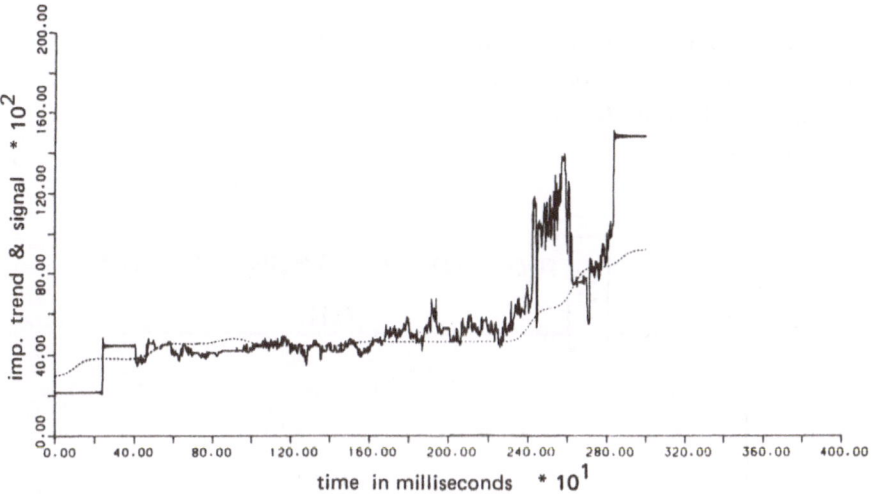

Fig. 8. Recovered acoustic impedance profile from the trend seismogram versus the well log impedance profile.

RESULTS ON TANK DATA

We have encouraging results from some data obtained in a water tank and we are therefore extremely grateful to Dr Neil Goulty, of the University of Durham, for providing the data.

Description of the data set

The two data sets were shot in a watertank. This has the advantage that we have a controlled measurement and that the outcome of the inversion result can be compared with the known situation. The source used in this experiment was known to emit an angle dependent signature, so first this angle dependency had to be removed (we did it in the wavenumber-frequency domain) to simulate point source data.

Therefore, one of the two datasets consisted of measurements of the source wavelets. The source in the experiment was a piezo-electric transducer. To calculate the wavefield of the piezo electric transducer in the wavenumber-frequency domain it is necessary to know the different wavelets emitted by the source at different angles. In Figure 9 the experimental setup is shown. The tank is completely filled with water and a receiver is fixed at the centre of the measured profile. The source is moved from one side of the profile to the other. For each position of the source the wavefield emitted is measured by the receiver. As the geometry of the experiment is known, the different angles of emittance corresponding to the different wavelets can be calculated. The parameters for this experiment were as follows:

Receiver depth	0.18	m
Source depth	0.005	m
Sampling interval	0.25E−6	s
Samples per trace	2048	

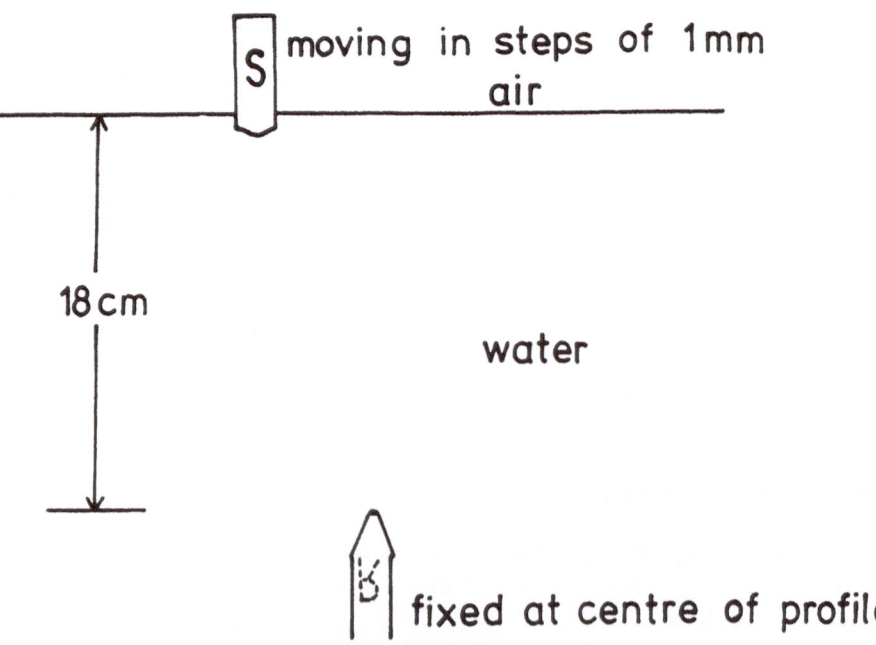

Fig. 9.   Experimental setup for measurement of the source directivity.

In total 361 measurements were made with 0.001 m between source positions. Measurement 181 was made with the source exactly above the receiver, with 180 measurements to each side. A plot of this dataset can be found in Figure 10. In Figure 11 a few wavelets have been enlarged; the ghost is included in these wavelets. It can be seen from this figure that the transducer emits a directionally dependent wavefield.

The second dataset consists of 48 CMP gathers. In Figure 12 the setup of this experiment is shown. The tank was filled with three materials; nilotron, perspex (both plastics) and water. The source-receiver pairs were located at the top of this configuration. The parameters for this experiment were as follows:

Depth of receiver	0.005 m
Depth of source	0.005 m
Sampling interval	0.25E−6 s
Samples per trace	2048
Fold of cover	48
CMP spacing	0.004 m
Near-trace offset	0.015 m
Offset increment	0.002 m
Far-trace offset	0.109 m
Length of profile	0.192 m

Inversion of the Watertank Data

Our inversion scheme was applied to every CMP gather after directional deconvolution in the wavenumber-frequency domain. First the data were transformed to the τ-p domain and there the deconvolution for the source wavelet was done.

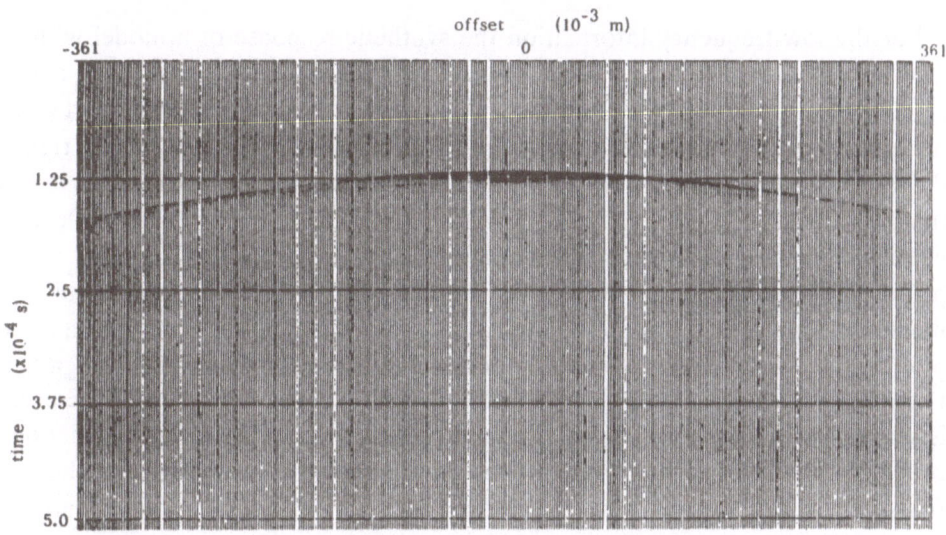

Fig. 10. Measurements of the source directivity, with the setup of Figure 9.

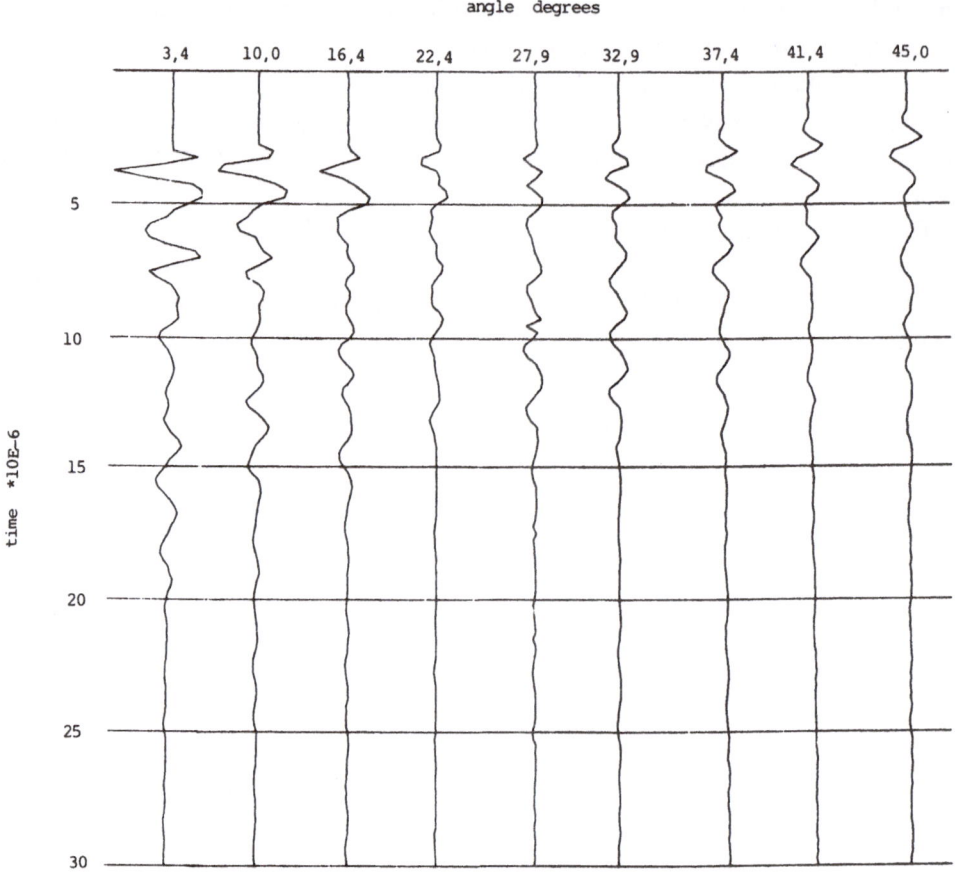

Fig. 11.   Enlargement of some of the measurements of the source directivity.

For the low-frequency information the synthetic response of a model with the true velocity and density was used. This means that to a certain extent the answer is fed into the data and we can expect that at least the low-frequency part of the reconstruction is correct. However, this synthetic model was only exactly right at the extreme left-hand side of the profile. In other words, we used only one low-frequency model for the entire line. After inversion of every plane wave component a move-out correction was applied to make it possible to stack all the plane wave results to one vertical incidence result. These stacked results are plotted in Figure 13 which shows the calculated profile and the true profile in one plot, and in figure 14 which shows the calculated profiles from all CMP gathers. From these figures it can be seen that the reconstruction of the admittance profile is satisfactory at the left-hand end of the profile, where the trend information is exact. Going to the right, the reconstruction becomes worse, since the low-frequency information is wrong. Since there is no low-frequency information in the data itself it cannot be expected that the data will ever change the low-

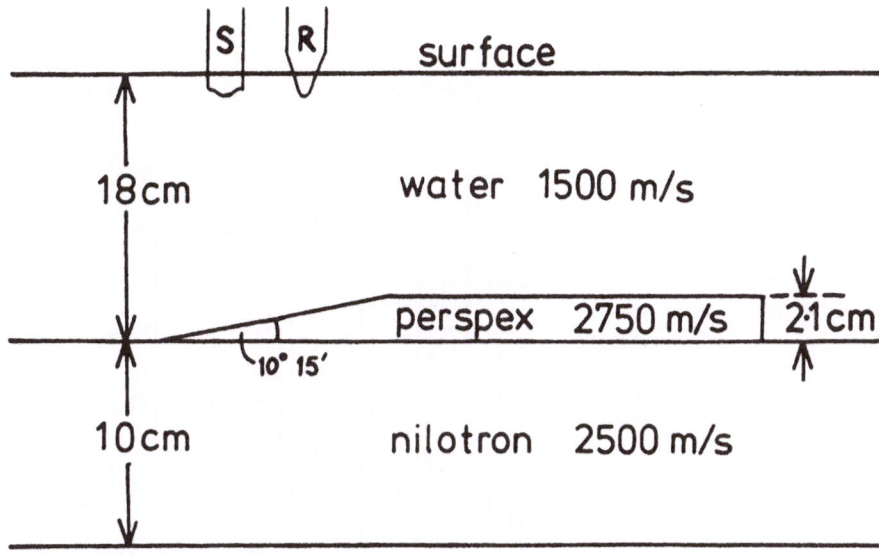

Fig. 12.   Experimental setup for the watertank data measurements.

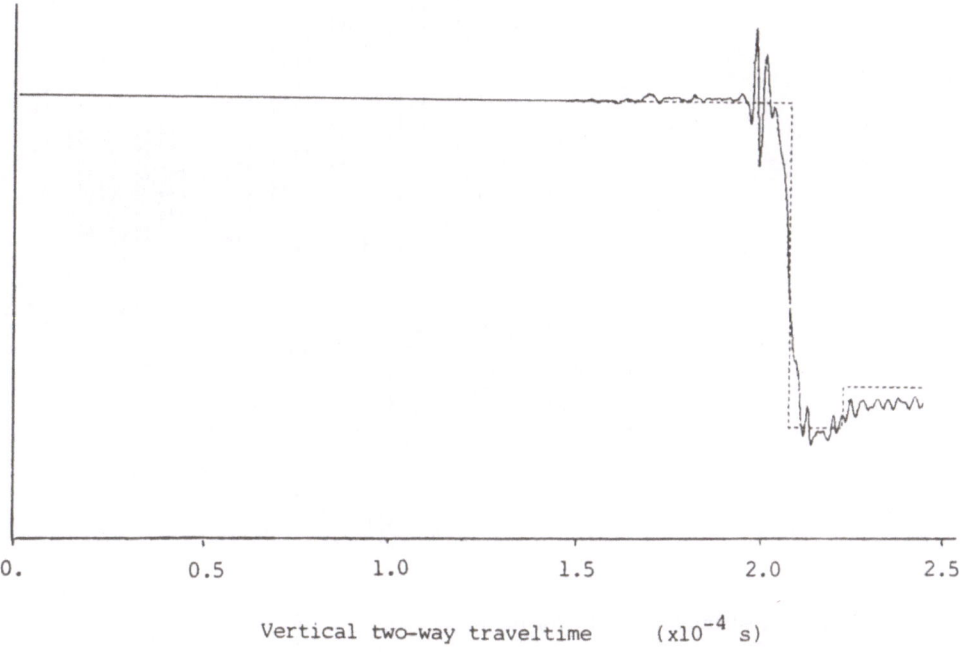

Fig. 13.   Admittance profile, solid line: inversion result; dashed line: true profile.

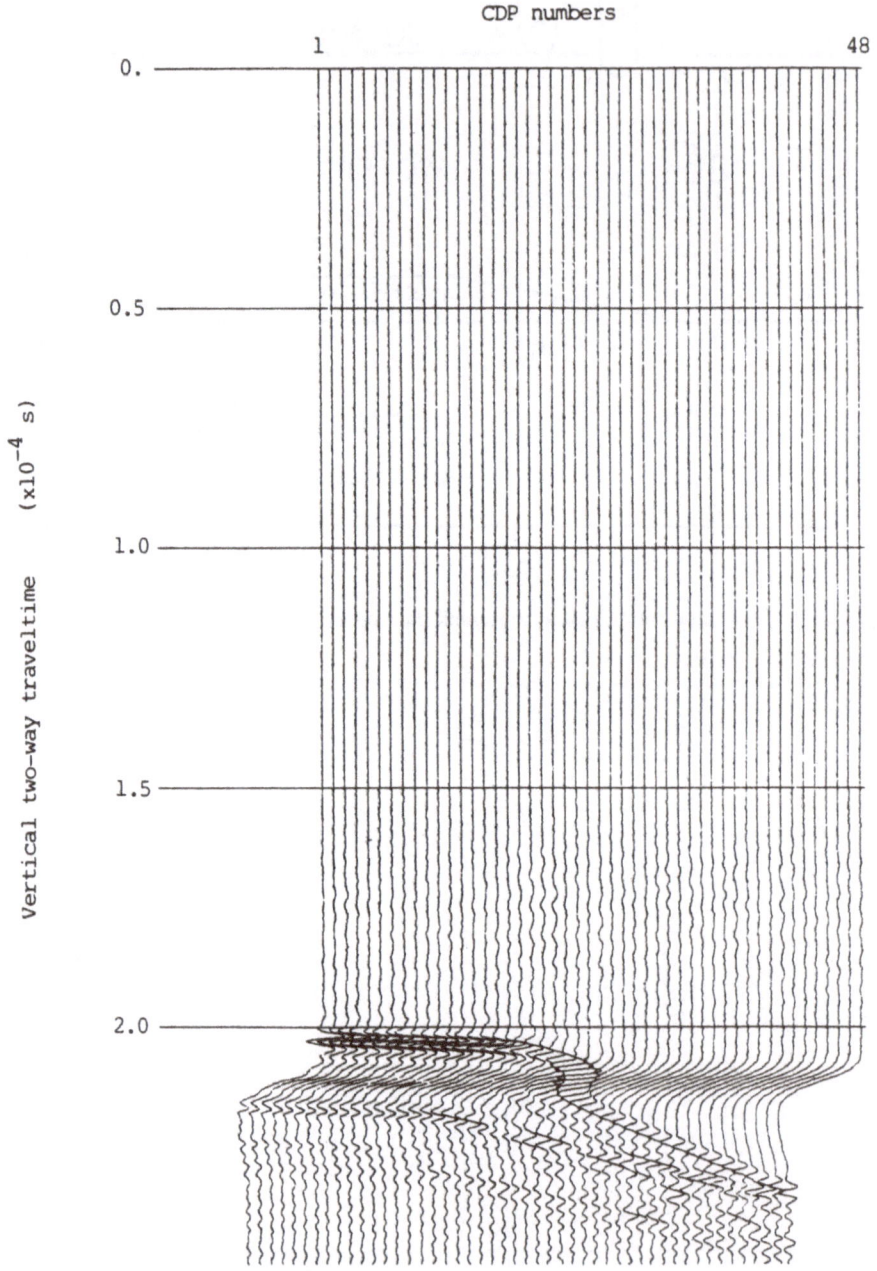

Fig. 14.   Recovered admittance profiles for the whole line.

frequency part of the reconstructed profile. In other words, for the low-frequency trend information you get out exactly what you put in.

INVERSION OF REAL SEISMIC DATA

As a final test of the inversion scheme we applied the scheme to a real marine seismic data set shot recently in the North Sea. These data consist of shot gathers of 208 traces with 2502 samples per trace and a sampling interval of 2 ms. The offset was 125 m and the distance between hydrophone groups was 12.5 m. Since these data were shot over a geological structure with almost no dip, it was decided to use a shot record instead of a CMP gather. (We have processed only one shot record.) The advantage of using a shot gather instead of a CMP gather is that the spatial sampling interval is much smaller.

In working with this data set we encountered all sorts of problems which are really obscuring the final inversion results:

(1) The wavefield emitted by the airgun array, used as the source in this data set, is unknown. We had to make an estimate of the wavefield using a modelling program.
(2) The instrument response is unknown, so we cannot convert the measured field values to actual pressure values.
(3) The given offset in the observer's logs and both the depth of the source and of the cable turned out to be wrong.
(4) Spatial aliasing problems made it necessary to resample the data to 8 ms.
(5) The experimental relation between velocities derived from a normal velocity analysis and densities as given by Gardner *et al.* (1974) did not provide us with a very reliable estimate of the densities.
(6) Since we do not yet know precisely how to cope with the effect of the free surface analytically, we had to make some approximations in doing so. First, the water bottom multiples were removed using predictive deconvolution. Secondly, we removed only the phase effect of the source and receiver ghost.
(7) Then the data had to be tapered in the offset direction for the large offset traces to cope with the so-called end-effect of the slant stacking program (the τ-p transform).
(8) For the scaling of the data in the τ-p domain we used a synthetic response calculated with the forward modelling scheme using a well-log as input geology model. The energy in the data and in the synthetic response were made equal in a certain time window before putting in the trend information.

The actual inversion was applied with both the time domain and the frequency domain versions of our inversion scheme. We present results for the inversion in the frequency domain when the trend is left out. The final result is then a primaries only reflection coefficients series. This can be compared with a similar series calculated from the log (Figure 15). Also we present results from the inversion in the time domain with trend information added to the data before

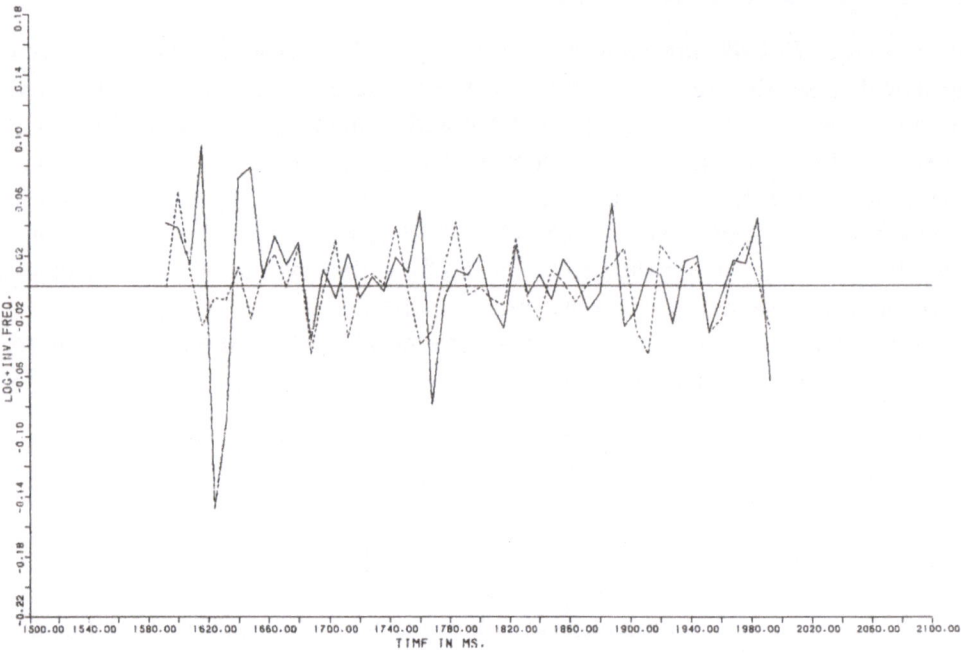

Fig. 15. Primary reflection series of the real data in a 500 ms time window. Dashed line: result after inversion in the frequency domain; solid line: series derived from well logs.

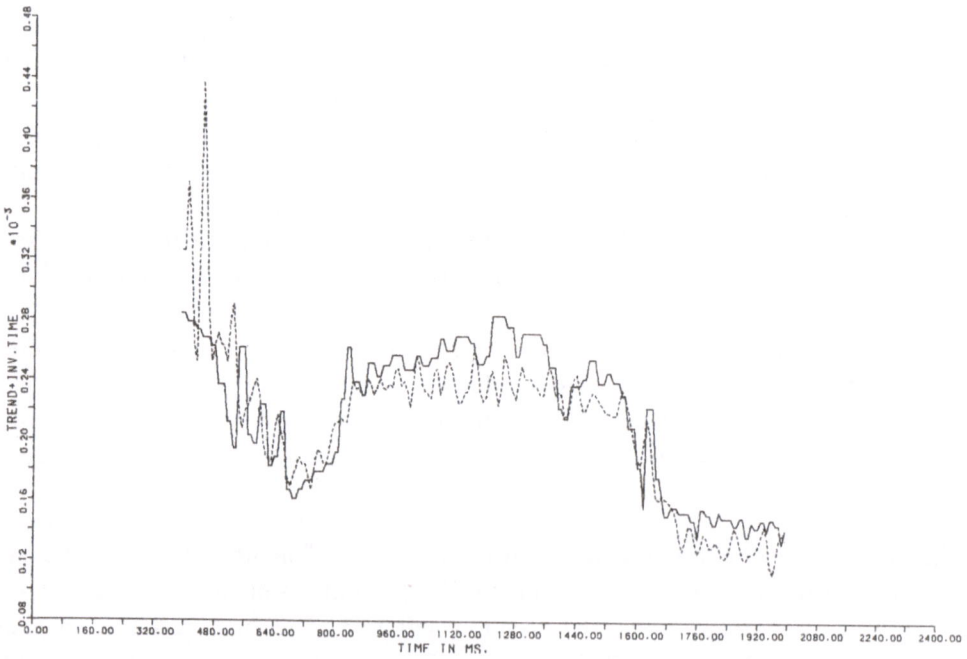

Fig. 16. Admittance profile for the inversion of the real data. Dashed line: result after inversion in the time domain; solid line: profile derived from well logs.

inversion. The result of this scheme can be compared with the admittance profile derived from the log (Figure 16).

In comparing our inversion results with the log, it can be seen that the fit is not everywhere as good as we hoped. There are a number of reasons why we think that the misfit is not entirely due to a failure of the inversion scheme. In our opinion the most important reason is the lack of knowledge about the acquisition parameters and absence of near-field measurements at the source for the determination of the source wavefield. Furthermore, it is always possible that the logs are not a real good model of the actual earth. The problems we have to reconcile seismic data and logs are of course not unique to us.

To have some sort of check on the performance of our inversion scheme on the real data we did the following 'round-trip':

(1) Derive from the admittance profiles the density and the velocity
(2) Use this density and velocity as the input for our forward modelling program.
(3) Put in water bottom multiples, source and receiver ghost and the source wavelet
(4) Transform the data set to the $x-t$ domain

This synthetic data set can then be compared with the original input real data. Both sets are plotted in Figures 17 and 18. We think that the agreement in the two sections is satisfactory, so we have confidence in our inversion scheme. The other check we made was to use the actual log in the same scheme described above. The synthetic data set calculated with the log as input is plotted in Figure 19. We think that the agreement between our inversion result and the original data is better than the result calculated with the log as input.

Conclusions

We have developed a one-pass scheme to recover the acoustic impedance profiles of the equivalent horizontally layered earth models from CMP data. This scheme is stable in the presence of noise.

It is crucial to know the source wavefield. If the source is not small, it must be collapsed to a point before the plane wave decomposition. That is, source directivity must be taken into account.

When the goal is to recover an admittance profile, trend information must be added to replace the missing low frequency information. This must be done with great care. The primary reflection coefficients series, however, can be recovered without adding trend information.

Since the trend information has a basically different frequency content than the seismic data, there is no hope that the data can ever influence the general shape of the admittance profile.

There are a number of points in our analysis that are open to discussion. For example: How valid is the assumption that we are dealing with an acoustic

Fig. 17.   Sythetic x–t section after forward modelling with the inversion results. Only third- and lower-order water bottom multiples included.

Fig. 18. $X-t$ section of the original data set.

Fig. 19.   Synthetic x–t section after forward modelling with the well log data as input.

horizontally layered earth? How reliable is predictive deconvolution to remove the water bottom multiples? How far can the misfit between the inversion result on the real data and the log be explained by errors in the log? Or, in other words, what is a better model of the earth: our inversion result or the well-log?

Improvements in the scheme could be made by extending it to locally-plane *dipping* reflectors, where it could then be applied to shot gathers instead of CMP gathers; by using an elastic earth model instead of an acoustic one; and by using a wave-theoretical approach to remove the effects of the free surface. These are topics we will consider in our future research programme.

References

Aki, K. and Richards, P. G., 1980, *Quantitative Seismology*, Freeman, San Francisco.

Backus, G. and Gilbert, F., 1970, Uniqueness in the inversion of inaccurate gross Earth data, *Philo. Trans. Roy. Soc.* **A266**, 123–192.

Berkhout, A. J., 1984, *Seismic Resolution*, Geophysical Press, London.

Berkhout, A. J., 1987, A frequency domain formulation of linearized inversion problems, *S.E.G. Expanded Abstracts, New Orleans, October 1987*, pp. 817–820.

Diebold, J. B. and Stoffa, P. L., 1981, The traveltime equation, τ-p mapping, and inversion of common-midpoint data, *Geophysics*, **46**, 238–254.

Du Cloux, 1986, Symmetry properties of elastodynamic wave fields and their application to space-time scattering theory, Report No. 1986–14, Faculty of Electrical Engineering, Delft University of Technology, Delft, The Netherlands.

Fokkema, J. T. and Ziolkowski, A., 1987, The critical reflection theorem, *Geophysics* **52**, 965–972.

Fuchs, K. and Muller, G., 1971, Computation of synthetic seismograms with the reflectivity method and comparison of observations. *Geophys. J. Roy. Astron. Soc.* **23**, 417–433.

Gardner, G., Gardner, L., and Gregory, A., 1974, Formation velocity and density — the diagnostic basis for stratigraphic traps, *Geophysics* **39**, 265–277.

Koefoed, O. and De Voogd, N., 1980, The linear properties of thin layers, with an application to synthetic seismograms over coal seams, *Geophysics* **45**, 1254–1268.

Mayne, W. H., 1962, Common reflection point horizontal stacking techniques, *Geophysics* **27**, 927–938.

Parkes, G. E., Ziolkowski, A., Hatton, L., and Haugland, T., 1984, The signature of an air gun array: computation from near field measurements including interactions — practical considerations, *Geophysics* **49**, 105–111.

Robinson, E. A. and Treitel, S., 1980, *Geophysical Data Analysis*, Prentice-Hall, Englewood Cliffs.

Schultz, P. S., Velocity inversion by wavefront synthesis, PhD thesis, Stanford Unversity.

Stoffa, P. L., 1985, Analysis and processing of wide-angle reflection and refraction seismic data in the τ-p domain, *in Advances in Geophysical Data Processing*, Vol. 2, JAI Press, pp. 81–117.

Stoffa, P. L., Diebold, J. B., and Buhl, P., 1981, Inversion of seismic data in the τ-p domain, *Geophys. Res. Lett.* **8**, 869–872.

Stoffa, P. L., Diebold, J. B., and Buhl, P., 1982, Velocity analysis for wide aperture common midpoint data, *Geophys. Prospect.* **30**, 25–27.

Tarantola, A., 1984, Inversion of seismic reflection data in the acoustic approximation, *Geophysics* **49**, 1259–1266.

Treitel, S., 1987, Invited introductory paper: Geophysical inversion: the dream and the reality, *S.E.G. 57th Annual Meeting, New Orleans, 11–15 October 1987, Expanded Abstracts*, p. 429.

Widess, M. B., 1973, How thin is a thin bed?, *Geophysics* **38**, 1176–1180.

Ziolkowski, A. M. and Fokkema, J. T., 1986, Tutorial, The progressive attenuation of high-frequency energy in seismic reflection data, *Geophys. Prospect.* **34**, 981–1001.

Ziolkowski, A. M., Parkes, G. E., Hatton, L., and Haugland, T., 1982, The signature of an air gun array: computation from near field measurements including interactions, *Geophysics* **47**, 1413–1421.

Delft University of Technology,
P.O. Box 5028,
2600 GA Delft,
The Netherlands

Index

MODERN APPROACHES IN GEOPHYSICS

formerly *Seismology and Exploration Geophysics*

1. E. I. Galperin, Vertical Seismic Profiling and Its Exploration Potential. 1985.
                                                          ISBN 90–277–1450–9.
2. E. I. Galperin, I. L. Nersesov and R. M. Galperina, Borehole Seismology. 1986.
                                                          ISBN 90–277–1967–5.
3. Jean-Pierre Cordier, Velocities in Reflection Seismology. 1985.
                                                          ISBN 90–277–2024–X.
4. Gregg Parkes and Les Hatton, The Marine Seismic Source. 1986.
                                                          ISBN 90–277–2228–5.
5. Guust Nolet (ed.), Seismic Tomography. 1987.   ISBN 90–277–2521–7.
6. N. J. Vlaar, G. Nolet, M. J. R. Wortel and S. A. P. L. Cloetingh (eds.), Mathematical
   Geophysics. 1988.   ISBN 90–277–2620–5.
7. J. Bonnin, M. Cara, A. Cisternas and R. Fantechi (eds.), Seismic Hazard in Mediter-
   ranean Regions. 1988.   ISBN 90–277–2779–1.
8. Paul L. Stoffa (ed.), Tau-p: A Plane Wave Approach to the Analysis of Seismic
   Data. 1989.   ISBN 0–7923–0038–6.
9. V. I. Keilis-Borok (ed.), Seismic Surface Waves in a Laterally Inhomogeneous
   Earth. 1989.   ISBN 0–7923–0044–0.